SpringerBriefs in Applied Sciences and Technology

Forensic and Medical Bioinformatics

Series editors

Amit Kumar, Hyderabad, India
Allam Appa Rao, Hyderabad, India

More information about this series at http://www.springer.com/series/11910

P.V. Lakshmi · Wengang Zhou · P. Satheesh
Editors

Computational Intelligence Techniques in Health Care

 Springer

Editors
P.V. Lakshmi
Department of Information Technology
GITAM Institute of Technology
GITAM University
Visakhapatnam, Andhra Pradesh
India

P. Satheesh
MVGR College
Vizianagaram, Andhra Pradesh
India

Wengang Zhou
DuPont Pioneer
Johnston, IA
USA

ISSN 2191-530X   ISSN 2191-5318   (electronic)
SpringerBriefs in Applied Sciences and Technology
ISSN 2196-8845   ISSN 2196-8853   (electronic)
SpringerBriefs in Forensic and Medical Bioinformatics
ISBN 978-981-10-0307-3   ISBN 978-981-10-0308-0   (eBook)
DOI 10.1007/978-981-10-0308-0

Library of Congress Control Number: 2016935565

Printed on acid-free paper

This Springer imprint is published by Springer Nature
The registered company is Springer Science+Business Media Singapore Pte Ltd.

Preface

This volume seeks to provide evidential research in emerging areas of computational intelligence techniques and tools, with a particular focus on emerging trends and applications in healthcare. Healthcare is a multifaceted domain, which incorporates advanced decision-making, remote monitoring, healthcare logistics, operational excellence, and modern information systems. In recent years, computational intelligence methods are being studied in order to address the scale and the complexity of the problems in the healthcare domain. Computational intelligence provides considerable promise for advancing many aspects of the healthcare practice, including clinical aspects as well as administrative and management aspects. This volume is a collection of various aspects of computational intelligence methods that are carried out in applications in different domains of healthcare.

Contents

Chapter 1
Bioinformatics, Genomics and Diabetes

Gumpeny Ramachandra Sridhar and Gumpeny Lakshmi

Abstract Bioinformatic analysis has been a key in unraveling the genetic basis of diabetes mellitus, which figured predominantly among target diseases for research after the human genome project. Despite extensive research the genetic contribution using current methods explains less than 10 % of predisposition. Data from next generation sequencing is bound to alter diagnosis, pathogenesis and treatment targets. Insight into the fine genetic architecture allows a fine grained classification of the diabetes spectrum, allowing primary preventive methods in at-risk individuals. In this quest the role of computational, statistical and pattern recognition would play increasingly major roles.

1.1 Introduction

Diabetes mellitus is a metabolic disorder with increasing prevalence the world over. It accounts for substantial disability death, economic and socioeconomic loss. It results from an imbalance between the need and availability in the body for insulin, a protein hormone secreted by β cells of pancreas. There is a complex interaction of genetic factors, environment and lifestyle in the expression of the disease. Extensive work on the genetic basis has generated enormous data, which in the current state explains for only a minor part of its cause. Next generation technologies applied to diabetes is providing insights to the elaborate checks and balances in normal physiology, whose disturbances result in susceptibility to and expression of diabetes. A number of hitherto unexplored regulatory factors such as regulatory RNA, epigenetic influences and microbiota found in the gut have come to the forefront in the cause and course of this disease.

G. Ramachandra Sridhar (✉) · G. Lakshmi
Endocrine and Diabetes Centre, 15-12-15 Krishnanagar, Visakhapatnam 530 002, India
e-mail: sridharvizag@gmail.com

© The Author(s) 2016
P.V. Lakshmi et al. (eds.), *Computational Intelligence Techniques in Health Care*, Springer Briefs in Forensic and Medical Bioinformatics, DOI 10.1007/978-981-10-0308-0_1

1

1.2 Background

Diabetes, with a worldwide prevalence of 382 million people in 2013 is projected to nearly double that figure by 2035 [1]. Arising from a genetic underpinning which interacts with environmental factors, current genetic technologies have identified many common variants which contribute to it. However these explain only a small fraction of diabetes heritability. Newer technologies can improve our genetic understanding of diabetes [2].

Diabetes mellitus is broadly classified into *Type 1 diabetes*, which often presents in the young as a result of pancreatic β cell loss. It is characterized by insulin deficiency and circulating autoimmune markers such as antibodies to glutamic acid dehydrogenase. The proportion of people with type 1 diabetes is less, ranging from 2 to 20 %. Heritability was explained up to 80 % by genetic factors, principally HLA class II alleles, other loci encompassing insulin gene, CTLA4, PTPN22, and interleukin 2 receptor [2].

Type 2 diabetes mellitus, which is more common, presents in adulthood, although obesity and sedentary lifestyle have seen children presenting with type 2 diabetes mellitus. The largest number of type 2 diabetes subjects are reported from China and from India [1].

Latent autoimmune diabetes in adults is also referred to as LADA. It has features of both type 1 and type 2 diabetes: GAD antibodies are found circulating in adults first diagnosed after the age of 35. They do not require insulin for glycemic control during the first six months of diagnosis. Since GAD antibodies are not routinely measured, a number of these cases are undiagnosed and clubbed under type 2 diabetes mellitus.

Maturity-onset-diabetes in the young or MODY is a monogenic form of diabetes. A number of genetic causes were identified, and the number is increasing. Individuals with MODY are diagnosed usually below 25 years. More than 200 mutations are described involving GCK (MODY2), HNF1A (MODY3) and PDX1 genes [3].

Maternally inherited diabetes and deafness (MIDD) results from a mutation in the mitochondrial DNA (A3242G mutation). Derived from the mother, it is transmitted maternally. Other neurological abnormalities may accompany diabetes and deafness.

Neonatal diabetes mellitus has been limited to diabetes with first onset at birth or before the first six months of birth. It may be transient or permanent. A number of genetic mutations were reported (KCJN11, SUR1, GCKm INS), which can be diagnosed only by genetic sequencing. A proper diagnosis is crucial, because they may respond to sulfonylurea drugs given orally, and may not require insulin injections [4].

Gestational diabetes mellitus is defined when diabetes is first identified during pregnancy. *Secondary forms of diabetes* result from other endocrine or pancreatic diseases.

1.3 Genetics and Heritability of Type 2 Diabetes

There has for long been evidence for genetic basis in subjects with type 2 diabetes: monozygotic twins have a nearly 70 % concordance of diabetes compared to dizygotic twins. Studies on the genetic basis for type 2 diabetes ranged from linkage studies to candidate gene studies, culminating in the genome wide association studies (GWAS); the last has to date provided the most extensive data, which, however is not yet translatable to clinical care [2, 5, 6].

Linkage analysis is used in the search for putative genes responsible for diabetes. A rough location of the gene is looked for, in relation to a DNA sequence called a genetic marker, which is another known sequence of DNA. Using this method, gene coding for the protein calpain 10 (CAPN10) was identified on chromosome 10. As a prelude to other loci to be identified, the calpain 10 was a protease which had no well known functions in the metabolism of glucose. However replications studies of the gene were largely unsuccessful. Variants in the TCF7L2 intronic variant (rs7903146) has been the most consistently replicated genetic association with type 2 diabetes [7].

Candidate gene studies. Greater success was obtained with PPARG gene, which was associated with type 2 diabetes mellitus. A number of studies have replicated the association. A variant, expressed in the adipose tissue showed increased transcription, improved insulin sensitivity and was protective against type 2 diabetes [8]. The other candidate which was identified was KCJN11 variant. It codes for the potassium-ATP channels, which are target for sulfonylurea group of antidiabetic drugs. Polymorphisms of ABCC8 gene, E23K polymorphisms in KCNJ11 and P12A in PPARG acted in a synergistic manner to increase the risk of type 2 diabetes mellitus [9].

Genome wide association studies (GWAS) scale up markers for genetic susceptibility with improved power and resolution. They are not yet employed in routine clinical care, and even if they would be, one must consider whether necessary time, financial and computational resources and expertise are available. In an ideal scenario, the genome sequencing consists of the 'complete base sequence for all chromosomes' in an individual.

In the shotgun sequencing method, DNA of interest is cut into small fragments randomly; next computer algorithms are employed to put back the sequence reads back into longer stretches, which needs adequate overlap obtained by deeper sequencing. The scaffolds, obtained after joining together of *contigs* from the initial assembly are in turn joined into linkage groups or are placed on chromosomes. Intensive computational methods are applied in analysis.

GWAS for diabetes are used to evaluate variability of genomes with susceptibility to diabetes using either data that is either population-based or family-based [10]. While GWAS identified a number of common polymorphisms that are found with complex traits, these explained only a small part of the disease expression, suggesting the existence of 'missing heritability' [11]. GWAS allowed unbiased interrogation of SNP linkages leading to better understanding of disease cause and identification of loci harboring disease associated variations. The theoretical ability to improve risk-prediction, clinical diagnosis and personalized treatment was not met [12]. But these studies were limited by the difficulty in interpreting the results and by the low proportion of heritability that could be accounted for by these markers.

In GWAS of diabetes mellitus, unbiased interrogation of millions of common variants associated with the disease, without restricting to known or suspected genes was carried out [11]. Two new loci related to diabetes were identified, viz HHEX and SLC40A8 [13] and were replicated in three separate reports [14–16].

Further, a meta-analysis comprising more than 50,000 subjects including both European and non-European cohorts was performed [17, 18]. Replication of findings from other ethnic groups increased the confidence in ascribing their association with diabetes. Replication of the following variants were identified: KCNQ1, UBE2E2, C2CD4A-C2CD4B, ANKI, GRK5, RASGRP1, PAX4, PPARγ, KCNJ11, TCF2, TCF7L2, CDKAL1, CDKN2A-CDKN2B, IDE-KIF11-HHEX, IGF2BP2 [2, 6].

A meta-analysis of GWAS (24,488 cases, 83,964 controls) was done among subjects with ancestry from Europe, East Asia, South Asia, Mexico and Mexico-America [19]. In this large cohort, there was an excess in directional consistency of T2D risk alleles. Seven novel T2D susceptibility loci were identified, which improved with fine-mapping resolution.

1.4   Next Generation Sequencing

Initial sequencing technologies, from the Sanger di-deoxy chain termination methods and its improvements required large quantities of DNA for analysis [20]. Manual Sanger sequencing method gave way to first generation automated DNA sequencers, leading to a faster and more accurate sequencing of the human genome. As of 2015, more than 150 variants for type 2 diabetes were mapped to over 120 loci, with more expected to be discovered [2]. Yet these accounted for but a small proportion of heritability of diabetes, implying that more dense sequencing of the DNA landscape is required, including the regions which do not code for proteins.

Next generation sequencing (NGS) methods reduced both the price and the time for sequencing the human genome [21]. NGS methods can sequence millions of small DNA fragments in parallel, which are pieced together by computational techniques. Sequencing can be performed of either the entire genome or of specific areas of interest [22]. These can capture a broader spectrum of variations than the Sanger method. A number of NGS platforms are available from Roche/454 life sciences, Illumina, Applied biosystems, and Pacific Biotechnology [20, 23]. Analysis of genomic, epigenomic, exomic, protein binding to target sequences and protein DNA interaction are all possible (Table 1.1). Newer generation of high-throughput-next generation sequencing technologies such as Helioscope^{TM} single molecule sequencer, Single molecule real time (SMRT^{TM}) sequencer, single molecule real time

Table 1.1 Next generation sequencing technologies (Adapted from [23])

Type of analysis	Type of sequencing
• Genomic	• Whole genome sequencing • Whole exome sequencing
• Epigenome	• Bisulfite sequencing • ChIP-seq • DNase-seq • FAIRE-seq • Hi-C • ChIA-PET
• Transcriptome	• RNA sequencing
• Interactome	• IVV-HiTSeq • Y2H-seq

(RNAP) sequence, Nanopore DNA sequencer, VisiGen Biotechnologies real time single molecule DNA sequencer platforms, are bound to generate vast sequencing databases that require advanced bioinformatics tools for analysis. Such ultra-deep sequencing methods will change the way in which one understands disease causes, new ways of diagnosis and treatment, using a single DNA molecule at a cost-effective price [24]. The use of exome sequencing or other methods ultimately depends on choosing the biomedical question that is sought to be answered and the resources that are available [25].

Current limitations for use in the clinical setting, which should be overcome by improved availability and lowered cost of technology, include acquisition of infrastructure, and more importantly, individuals who can analyse, annotate and draw out the clinically relevant information from the data [22]. Integration is also required not only for interpretation, but also involves billing and insurance, consent process and counseling based on the available information [21].

1.4.1 Methodological Aspects

Large quantity of data generated from NGS technologies needs methods for annotation, curation and analytical bioinformatics methods [6].

Conventional methods and NGS can be combined to identify mutations. In families where a complex condition affects many individuals, rare alleles with high penetrance may be present. While linkage studies were unable to identify many loci in diabetes, a combination with exome or genome sequencing in family pedigrees may be employed to identify penetrant mutations that are specific to a particular family [26]. Similarly while admixture mapping and association tests were independently applied to identify genes of complex diseases, a combination of these approaches may also be employed. Utilizing data from Genetic Analysis Workshop 18 [27], combining admixture and association approaches was shown to be promising to detect quantitative trait loci, by increasing the detection power using two separate sources of genetic signals [28]. When population-based and family-based GWA data are combined, the requirement for multiple testing may be reduced. Such a method was employed to rank SNPs related to diabetic nephropathy where unrelated singleton and parent-offspring trio cases and controls were studied. This combined method was useful to identify associations that are otherwise not easy to detect in the presence of multiple testing corrections [10]. Identification of rare variants following NGS is an important aspect. Statistical tools are employed to pool or collapse rare single nucleotide variants to enhance the effect. Where such trait is expressed in different directions, the signal is weakened. In situations such as these, a vector support machine (BSVM: backward support vector machine) may be utilized to identify and select the variants [29]. The nonparametric variant selection accounts for confounding factors by weighting or collapsing according to the positive or negative influence.

As NGS often gives base calls with low quality scores, preprocessing methods are often employed to improve the accuracy. Commonly employed methods are *masking*, which substitutes low quality base calls with underdetermined bases and *trimming* which removes low quality bases in shorter read lengths. Yun and Yun [30] showed that of the two methods, masking was the more effective in reducing false positive rates in identifying SNPS [30].

1.4.2 Diagnosis of Diabetes

Although diabetes is classified as type 1 diabetes, type 2 diabetes etc., the spectrum is continuous [31]. NGS genetic analysis is most likely to improve the diagnosis of ***monogenic diabetes*** which results from defects in single genes, leading to dysfunction and loss of β cells [32]. NGS is expected to improve diagnosis in a cost effective manner which helps both in understanding the pathogenic mechanism as well as to decide the appropriate treatment. A combination of NGS and genomic region enrichment, called the 'target-region capture sequencing' has been shown to be a useful method in identifying monogenic diabetes [33]. NGS method applied to monogenic diabetes was shown to be sensitive in simultaneously analyzing all known genes leading to monogenic diabetes, and could increase the detection rate of mutations in conditions such as neonatal diabetes, MODY, maternally inherited diabetes and deafness and familial partial lipodystrophy [34].

Newer sequencing technologies could also clarify what was considered a monogenic form of diabetes based on phenotypic features may not be so. Ketosis-prone diabetes (KPD), with a predisposition to develop diabetic ketoacidosis is classified into A-β-subgroup when β cells are absent without islet cell autoimmunity and a strong family history of type 2 diabetes. In a study of 37 subjects from a follow up cohort of KPD patients, more than 70 % did not have causal polymorphisms of seven putative genes either in the proximal promoter or the coding regions (HNF4A, GCK, HNF1A, PDX1, HNF1B, NEUROD1 and PAX4). NGS recognized that KPD may not be a monogenic form of diabetes although low-frequency variants in HNF1A, PDX1 or PAX4 genes may play a pathogenic role in the disease [35].

When obesity co-exists with monogenic forms of diabetes, an accurate diagnosis helps in clinical care and in genetic counseling. NGS with Illumina HiSeq2000 was employed on PCR enriched microdroplets to characterize the molecular diagnosis of monogenic diabetes or obesity. This method could identify more than 95 % of gene mutations causing the phenotype [36]. Widespread availability and lowered costs could bring the technique into routine clinical care. Case reports of NGS based genetic techniques were published in the diagnosis of familial partial lipodystrophy, where mutational analysis of LMNA gene showed hot spot missense mutation (c.1444C > T, p.Arg482Trp) [37].

Type 1 diabetes, due to autoimmune destruction of the β cells of the pancreas is principally associated with Class II HLA genes (HLA-DRB1, HLA-DQB1) and class I genes (HLA-A, HLA-B), while the relation of polymorphisms at HLA-DRB3, HRB4, HLA-DRB5 are not well known. Utilization of NGS technology identified that DRB3*02:02 on the DRB1*03:01 haplotype could predispose to the risk of type 1 diabetes mellitus [38]. Viruses are implicated in the causation of type 1 diabetes mellitus, leading to autoimmune destruction of pancreatic β cells. To analyze whether viruses can be identified in the plasma of children with infection who went on to develop type 1 diabetes, next-generation sequencing for the viruses was carried out. But, viremia was not shown to precede seroconversion in young children with rapid onset type 1 diabetes [39]. Similarly a case-control study utilised next-generation sequencing methods to evaluate gut virome at the onset of autoimmunity of islet cells. In a small cohort of 19 cases developing type 1 diabetes, the stool virome did not show any dramatic alterations just preceding islet autoimmune injury [40].

Type 2 diabetes Type 2 diabetes has been the most studied for genetic underpinnings, which identified 65 common genetic variants that were recently employed to predict the onset of diabetes. Since a lower number of genetic variants did not predict onset of diabetes any better than clinical and anthropological measures, the authors increased the number of genetic variants to study whether a larger gene pool improves the predictive power. When used alone, the genetic score alone identified 19.9 % of incident cases of type 2 diabetes mellitus in contrast to the Framingham risk model which identified 30.7 %. However when the two were combined the detection rate increased to 37.3 % [41]. While exome sequencing was employed for rare T2DM variants having large effect, there were no identifiable variants in case-control setting, while exome sequencing in a family identified variation in the gene which encoded early endosome antigen 1 (cSNP in EEA1) [42].

1.4.3   Functions of Diabetes Associated Genes

Despite the identification of a large number of SNPs associated with the risk of type 2 diabetes mellitus, the functional significance of most, or their putative role in causation are unknown. A few have been understood such as MTNR1B, IRS1, TCF71.2, GIPR and a few others [2]. MTNR1B (melatonin receptor B) is related to melatonin, a chronobiotic factor distributed in the pancreas. It has been proposed to decrease cAMP/cGMP levels and impair insulin secretion. Insulin receptor substrate (IRS1) mediates insulin action through the cell membrane. Variant form is associated with reduced insulin induced activity of enzymes involved in skeletal muscle [43]. The transcription factor TCF7L2, which has a key function in Wnt signaling pathway results in lowered insulin secretion. It has other roles in influencing pancreatic β cell mass, maintenance and function [44].

Other forms of diabetes and NGS A GWAS of *gestational diabetes mellitus* identified genetic variants in CDKAL1 and MTNR1β were associated with risk of GDM. Employment of NGS would enhance an understanding of the pathogenetic aspects [45]. Identification of *monogenic forms of diabetes* would be possible from NGS technique, which enables sensitive simultaneous analysis of all genes [34]. *Neonatal diabetes mellitus,* can also be accurately diagnosed by whole exome sequencing. Identification of KCNJ11, ABCC8 and INS variants by NGS makes it possible to rapidly and cost-effectively diagnose and provide specific treatment [46]. Rare genetic disorders such as difficult to diagnose mitochondrial diseases can be identified by NGS technology. A case report was published where targeted exome sequencing identified a subject with diabetes mellitus, brain atrophy, autonomic neuropathy, optic nerve atrophy and amnesia. mtDNA deletions were associated with non-nutritional thiamine deficiency. A homozygous c.1672C > T (p.R558C) missense mutation in exon 8 of WFS1 was identified [47]. Genetic testing and identification of familial partial lipodystrophy is also described using NGS [37].

1.4.4   Pathogenesis of Diabetes

Diabetic nephropathy Understanding of the pathogenic processes of microvascular complications of diabetes is possible. End-stage-renal disease often results from diabetes mellitus. In a OVE26 murine model of diabetes, NGS-based transcriptome analysis of renal gene expression was performed. Compared to controls, in diabetes 638 genes were dysregulated [48]. More than 70 % were downregulated. Most genes were related to oxidative stress, endoplasmic reticulum stress or protein folding. Use of antihypertensive drug losartan altered gene expression. Such identification of individual genes suggest newer pathways for disease progression and consequently potential targets for treatment [48]. In a rat model of diabetic nephropathy, NGS was performed to study the changes in diabetic nephropathy that led to progressive chronic kidney disease. Obese diabetic ZS rat was employed as a model, and RNA-seq was performed. Multiple interactions of gene changes were observed, which were integrated into networks [49]. These led to renal inflammation and apoptosis. Multiple and redundant circuits were linked to renal fibrosis and destruction. Therefore NGS provides data that can be used in network models to correct dysregulation by novel treatment methods. Fibrotic signals were identified in diabetic nephropathy via transforming growth factor-beta (TGF-β). Transcriptional profile of renal tubular epithelial cells were studied after stimulation with TGF-β1 with RNA-Seq. 2027 genes were differentially expressed, in which TGF-β1 driven pro-fibrotic signal in epithelial cells of kidney in humans was shown to occur [50]. In humans pedigree studies using NGS models are being used to identify rare variants [51].

Diabetic retinopathy RNA-seq profiles in diabetic retinopathy identified transcript signature alterations early in diabetic retinopathy. They coded proteins involved in processes such as inflammation, formation of microvessels, apoptosis, WNT signaling, biology of photoreceptors, xenobiotic metabolism and glucose metabolism [52]. Upregulated crystalline transcripts seen in diabetes were inhibited with inhibitors of either RAGE or p38 MAP kinase. RNA-seq thus could identify transcripts that are specific for diseases and provide leads for development of drugs to target the pathological changes.

Epigenetics and NGS When it became apparent that the gene containing regions of the chromosome and their variants could not alone account for all of the phenotypic variability, the old concept of 'epigenesis' came to the centre-stage and is now a key target in NGS technology. Epigenetics refers to the study of DNA changes, post-translational changes of protein constituents of chromatin [53]. Epigenetic regulation can silence gene expression and acts as a signal between the environment and expression of genes. Well studied epigenetic phenomena include *DNA methylation*, which is a stable epigenetic signal that is transmitted across generations. A methyl group is covalently bound to cytosine in DNA at the CpG regions which are located in relation to gene regulatory regions. Hypomethylation generally increases the expression of genes, whereas hypermethylation generally inhibits gene expression. The other well studied phenomenon is *histone modification* in which gene expression regulators, the histones are post-translationally altered to regulate gene expression by modifying the exposure of DNA sequence accessibility. *RNA related* epigenetic phenomenon occurs via non coding RNAs or ncRNAs. RNA interference or antisense transcripts can regulate turning on or turning off of genes. These can occur at the level of the chromatin or at the post-transcriptional level [54].

The epigenetic regulation which can be considered complimentary to that encoded by DNA has been referred to as a 'dynamic and rapidly advancing research field' which would be impacted by emerging NGS technologies [55].

1.4.5 Epigenetics and Pathogenesis of Diabetes and Vascular Complications

Evidence is accumulating for epigenetic alterations contributing to the pathogenesis of metabolic syndrome, type 2 diabetes mellitus and the vascular complications [56, 57].

Epigenetic phenomenon were identified in the concept of fetal onset of adult diseases in which nutritional stress during pregnancy was associated with adverse metabolic conditions such as insulin resistance, coronary artery disease and hypertension. When animals were fed diet which was low in folate and vitamin B_{12}, their offspring showed altered methylation of CpG islands which was associated with obesity, insulin resistance and raised blood pressure [53]. In babies born of mothers with dysglycemia, epigenetic changes may be responsible for abnormal organ formation. Altered Acetyl-Co-A in the nucleus due to hyperglycemia may alter histone acetylation and methylation of DNA. In addition miRNA networks are important regulators for the proper development of pancreatic β cells. miR-7a2 was shown to negatively regulate insulin secretion of β cells in response to glucose [58].

To gain insight into the link between pathogenesis of diabetes and epigenetic changes, a data mining method was employed. A MEDLINE record search showed potential items linked with type 2 diabetes mellitus showed methylation and chromatin among the items ranked in the top-five terms [59]. On the basis of the observations, a hypothesis was proposed that short-chain fatty acids could affect DNA methylation and consequently alter the regulation of pro-inflammatory cytokines. Role of miRNA has been reported in β-cell regeneration; epigenetic changes can also influence the insulin gene expression by regulating insulin promoter via DNA methylation.

Clinical evidence has shown that achievement of euglycemia during the early stages of diabetes is associated with better clinical outcomes, even though continued levels of normal glucose were no longer maintained. This phenomenon is called *metabolic memory* and epigenetic changes were postulated as key factors. In a sample of subjects from Diabetes Control and Complications Study and from United Kingdom Prospective Diabetes Study, epigenetic alterations were studied in white blood cells. When blindly studied for histone activating and repression marks followed by promoter tiling analysis, subjects who were not in the intensive treatment group had enrichment of H3K9Ac loci. Hyperacetylated regions contained inflammatory genes STAT1, TNF alpha and IL1A, suggesting long-lasting epigenetic effects in those with poorer glycemic control at the beginning of the study [60]. A recent study showed that glucose induced changes in mir-125b and miR-1465p could activate NF-kB pathway and contribute to metabolic memory [61].

Gut microbiota, metabolic syndrome and diabetes Gut microbiota is the new player in the etiology of metabolic syndrome and diabetes. Newer sequencing technologies enabled identification of different microbial and viral species using their nucleotide sequences. Each human being has about 1.5 kg of bacteria, mostly located in the distal intestines. The combined genomes of the different gut microbes and the human host is called the *metagenome*. There is a complex interaction between the microbes and human host physiology and pathological processes. Newer sequencing technologies have increased the number of colonizing bacterial species to more than 35,000. Whole genome sequencing and use of 15 rDNA sequences are denatured, subjected to temperature gradient and digested. Utilizing whole genome sequencing, bioinformatics techniques identify functional capability by aligning with reference genomes [62]. Gut flora regulate energy balance by extracting energy from the diet, regulating gut hormone secretion such as peptide YY and by breaking down plant carbohydrates which are then absorbed in the human intestines [63]. Mutual interactions take place at the level of nucleic acids as well as of proteins [64]. With increasing access to NGS technology, initial animal and later human studies have shown differential composition of species among the lean, obese and those with diabetes mellitus. Interventions to correct the metabolic abnormalities reverted the species composition [65]. Metagenomic profiles devised from mathematical models allowed identification of subjects with diabetes-like metabolism. Targeting species identified by NGS makes it possible to catalog human microbiome, improve our understanding of their role [62] and finally devise methods to correct the metabolic disorders [66].

1.5 Issues, Controversies, Problems

Promising as newer technologies are in dealing with diabetes mellitus, there are issues that need to be cleared before the full potential can be reached. Cost of processing power, which had once been a significant problem is rapidly decreasing. It should be only a matter of time before the cost breaches the level when sequencing could be a useful and appropriate diagnostic and prognostic tool in routine clinical care. While costs come down, problems remain in the analysis and interpretation of the large quantities of data generated. Bioinformatic methods to synthesize the data into actionable results must keep pace with increasing computing power.

The way to distinguish normal from abnormal remains a significant bottleneck. With deep sequencing churning out more data, one needs a matching phenotypic characterization. The reason why currently available variants explain only a small percent of heritability could be because the clinical diagnoses are coarse, and do not match the precision of sequencing techniques. Conditions such as obesity, diabetes, hypertension or coronary artery disease are grouped together, and they really comprise a number of separate conditions.

To match phenotype with genotype requires characterization of clinical features, biochemical results, demographic, social and psychological parameters, and how they vary over time. It would then be possible to devise methods to latch phenotype with genotype in a clinically useful manner. Issues of privacy and ethics would necessarily have to be assessed, balancing benefits and harms [67]. Conditions that are identified must have high penetrance and must be actionable. Unlike BRCA mutations in families with carcinoma of breast, or some cases of familial hypercholesterolemia, diabetes is presently a more coarsely defined condition, where available clinical scores are as good and often better than genetic scoring system when used alone. The lack of sensitivity of genetic scores over clinical scores does not negate the value of the latter, but only suggests more appropriate refinement of genetic scoring system is required. Results from different ethnic groups, environmental influences, including epigenetic and metagenomic data would enhance the predictive ability.

Some of the known limitations of genetic scores are an insufficient size of genetic loci, poor ability of score to discriminate, with lesser incremental value when added to clinical risk, unknown clinical relevance for variants identified, and an inability to account for gene-gene, gene-environmental interactions [5]. Potential application of genetic studies are, ability to subtype the diabetes phenotype, identification of low-frequency and rare variants with large effect sizes using NGS, and development of newer statistical methods to study these wide-ranging variables [5].

1.6   Conclusion

Methodological, ethical, technological advances have the potential to translate risk loci into druggable targets: Carriers of the adrenergic receptor α2 variant (ADRA2A rs5536668) have decreased insulin secretion; agonist of α_{2a}AR yohimbine can increase the release of insulin in vitro. Blocking the receptor can increase insulin secretion in subjects carrying the risk allele [68]. Bioinformatics analysis suggested that plants contain nucleotide sequences similar to that of human insulin [69]. Studies in plant species using NGS could be a tool to identify transcriptomal molecular signatures relevant to plant tissue physiological roles [70]. Sequencing leaf transcriptome of C. pictus, transcripts related to anti-diabetic properties of C. pictus were identified. These were related to pathways of bixin biosynthesis and geraniol and geranial biosynthesis as major transcripts [70].

A systems biological approach to gene sequencing including protein analysis and phenotype-genotype association along with metabolomics can help in establishing links between SNPs and metabolic pathways, which can be identified using network based approaches. This aids in obtaining a holistic picture of the disease and therefore suggests pathways where intervention is possible. NGS could thereby open up the goal of personalized medicine in diabetes, which matches treatment based on genetic, phenotypic and psychosocial characters, to improve clinical outcome and reduce unnecessary side effects for the individual [71, 72]. Pointers towards such personalization is shown by genetic markers being used to identify the risk of developing diabetes and lipid variants [73]. Understanding such interactions allows for development of novel targets for both lipid disorders and diabetes [74]. Curation of databases containing data for estimation of genetic risk are being put into place which could aid in the progress of personalized medicine [75, 76].

Personalized medicine is a step behind precision medicine, which uses a comprehensive view of the patient's condition and its time trajectory. A greater individualization of treatment is thus possible, or precision medicine [77]. Integration and affordability would form a critical component, as would education of health care providers in utilizing the flood of data.

Ultimately physicians should be able to provide genome-guided therapy at the point of care utilizing pharmacogenomics data and clinical decision support which are all integrated into electronic medical systems [2, 78].

Data could be generated faster than it can be analyzed, understood and put to use. When the ability to utilize information matches the rate at which it is being generated, the promise envisaged by the HGP at the turn of the new millennium could be realized. Understanding the etiology of diabetes to its management stands to be altered. Diabetes would be more precisely defined and classified with a genetic basis for the clinical presentation. This would enable a pathogenesis based treatment besides allowing preventive measures to be provided even before the disease sets in. Combining genetic data obtained by deep sequencing with clinical criteria would allow a more precise allocation of resources for primary prevention (lifestyle, diet, exercise and where relevant, specific medicines). Insight into pathogenesis of complications could also allow methods to target specific drugs which would be effective, without significant complications. Emerging factors such as privacy, security, ethical aspects and affordability would have to be tackled. Along with the generation of data, health care providers must be trained to deal with it, and interpret it so that the potential benefits can be seen. This brings the issue of affordability and equitable available of treatment. Issues unique to such technologies—data protection, and influence of political policy decisions must also be taken into account [79, 80]. While the costs of sequencing with NGS would come down, a bottleneck of interpreting the data would need to be sorted out [81]. Similarly the health delivery system is likely to face uncertainty because of the paradigm shifts new ways of conceptualizing and delivering treatment would emerge. Even though spirited discussions populate what constitutes a bioinformatician [82, 83] and findings of whole exome sequencing results do not yet fully correlate with medical findings [84], translation of science to the clinic is a work in progress, but a progress that is likely to evolve sooner than later.

References

1. Diabetes Atlas (2014) Available online at www.Idf.Org/diabetesatlas
2. Prasad RB, Groop L (2015) Genetics of type 2 diabetes—pitfalls and possibilities. Genes 6:87–123
3. Murphy R, Ellard S, Hattersley AT (2008) Clinical implications of molecular genetic classification of monogenic beta-cell diabetes. Nat Clin Pract Endocrinol Metab 4:200–213
4. Groop L (2015) Genetics and neonatal diabetes: towards precision medicine. Lancet 386:934–935

5. Lyssenko V, Lupi R, Marchetti P, Del Guerra S, Orho-Melander M, Almgren P, Sjogren M, Ling C, Eriksson KF, Lethagen AL et al (2007) Mechanisms by which common variants in the TCF7L2 gene increase risk of type 2 diabetes. J Clin Invest 117:2155–2163
6. Sridhar GR, Duggirala R, Padmanabhan S (2013) Emerging face of genetics, genomics and diabetes. Int J Diab Devel Countries 33:183–185
7. Tong Y, Lin Y, Zhang Y, Yang J, Zhang Y, Liu H, Zhang B (2009) Association between TCF71.2 gene polymorphisms and susceptibility to type 2 diabetes mellitus: a large human genome epidemiology (huge) review and meta-analysis. BMC Med Genet 10:e15
8. Majithiaa AR, Flannicka J, Shahiniana P, Guod M, Braya M-A, Fontanillasa P, Gabriela SB, GoT2D Consortium, NHGRI JHS/FHS Allelic Spectrum Project, SIGMA T2D Consortium 2, T2D-GENES Consortium, Rosenc ED, Altshuler D (2014) Rare variants in PPARG with decreased activity in adipocyte differentiation are associated with increased risk of type 2 diabetes. PNAS 111:13127–13132
9. Chavali S, Mahajan A, Tabassum R, Dwivedi OP, Chauhan G, Ghosh S, Tandon N, Bharadwaj D (2011) Association of variants in genes involved in pancreatic β-cell development and function with type 2 diabetes in North Indians. J Human Gen 56:695–700
10. Estus JL, Family Investigation of Nephropathy and Diabetes Research Group, Fardo DW (2013) Combining genetic association study designs: a GWAS case study. Front Genet 4:186. doi:10.3389/fgene.2013.00186. eCollections 2013
11. Chen Q, Sun F (2013) A unified approach for allele frequency estimation, SNP detection and association studies based on pooled sequencing data using EM algorithms. BMC Genomics 14 (Supplement 1):S1. doi:10.1186/1471-2164-14-S1-S1
12. Wang Q, Lu Q, Zhao H (2015) A review of study designs and statistical methods for genomic epidemiology studies using next generation sequencing. Front Genet 6:149. doi:10.3389/fgene.2015.00149
13. Sladek R, Rocheleau G, Rung J, Dina C, Shen L, Serre D, Boutin P, Vincent D, Belisle A et al (2007) A genome-wide association study identifies novel risk loci for type 2 diabetes. Nature 445:881–885
14. Diabetes Genetics Initiative of Broad Institute of Harvard and MIT, Lund University, Novartis Institutes of BioMedical Research, Saxena R, Voight BF, Lyssenko V, Burtt NP, de Bakker PI, Chen H et al (2007) Genome-wide association analysis identifies loci for type 2 diabetes and triglyceride levels. Science 316:1331–1336
15. Scott LJ, Mohlke KL, Bonnycastle LL, Willer CJ, Li Y, Duren WI, Eridos MR, Stringham HM, Chines PS et al (2007) A genome-wide association analysis of type 2 diabetes in Finns detects multiple susceptibility variants. Science 316:1341–1345
16. The Wellcome Trust Case Control Consortium (2007) Genome-wide association study of 14,000 cases of seven common diseases and 3000 shared controls. Nature 447:661–678
17. Voight BF, Scott LJ, Steinthorsdottir V, Morris AP, Dina C, Welch RP, Zeggini E, Huth C, Aulchenko YS et al (2010) Twelve type 2 diabetes susceptibility loci identified through large-scale association analysis. Nat Genet 42:579–589
18. Morris AP, Voight BF, Eslovich TM, Ferreria T, Segre AV, Steinthorsdottir V, Strawbridge RJ, Khan H et al (2012) Large-scale association analysis provides insights into the genetic architecture and pathophysiology of type 2 diabetes. Nat Genet 44:981–990
19. DIAbetes Genetics Replication and Meta-analysis (DIAGRAM) Consortium, Asian Genetic Epidemiology Network Type 2 Diabetes (AGEN-T2D) Consortium, South Asian Type 2 Diabetes (SAT2D) Consortium, Mexican American Type 2 Diabetes (MAT2D) Consortium, Type 2 Diabetes Genetic Exploration by Next-generation sequencing in multi-Ethnic Samples (T2D-GENES) Consortium (2014) Genome-wide trans-ancestry meta-analysis provides insight into the genetic architecture of type 2 diabetes susceptibility. Nat Genet 46:234–244
20. Tipu HN, Shabbir A (2015) Evolution of DNA sequencing. J Coll Phys Surg Pak 25:210–215
21. Durmaz AA, Karaca E, Demkow U, Toruner G, Schoumans J, Cogulu O (2015) Evolution of genetic techniques: past, present, and beyond. BioMed Res Int Article id: 461524. http://doi.org/10.1155/2015/461524

22. Behjati S, Tarpey PS (2013) What is next generation sequencing? Arch Dis Child Pract Educ 98:236–238
23. Ohashi H, Hasegawa M, Wakimoto K, Sato EM (2015) Next-generation technologies for multiomics approaches including interactome sequencing. BioMed Res Int Article id: 104209. http://dx.doi.org/10.1155.2015/104209
24. Pareek CS, Smoczynski R, Tretyn A (2011) Sequencing technologies and genome sequencing. J Appl Genet 52:413–435
25. Biesecker LG, Shianna KV, Mullikin JC (2011) Exome sequencing: the expert view. Genome Biol 12:128
26. Morris JA, Barrett JC (2012) Olorin: combining gene flow with exome sequencing in large family studies of complex disease. Bioinformatics 28:3320–3321
27. Bickeboller H, Bailey JN, Beyene J, Cantor RM, Cordell HJ, Culverhouse RC, Engelman CD, Fardo DW, Ghosh S, Konig IR et al (2014) Genetic analysis workshop 18: methods and strategies for analyzing human sequence and phenotype data in members of extended pedigrees. BMC Proc 8(Suppl 1):S1. doi:10.1186/1753-6561-8-S1-S1. eCollection
28. Yorgov D, Edwards KL, Santorico SA (2014) Use of admixture and association for detection of quantitative trait loci in the Type 2 Diabetes Genetic Exploration by Next-Generation Sequencing in Ethnic Samples (T2D-GENES) study. BMC Proc 8(Suppl 1):S6. doi:10.1186/1753-6561-8-S1-S6. eCollection
29. Fang YH, Chiu YF (2013) A novel support vector machine-based approach for rare variation detection. PLoS ONE 8(8):e71114. doi:10.1371/journal.pone.0071114
30. Yun S, Yun S (2014) Masking as an effective quality control method for next-generation sequencing data analysis. BMC Bioinform 15:382. doi:10.1186/s12859-014-0382-2
31. Donath MY, Ehses JA (2006) Type 1, type 1.5, and type 2 diabetes: NOD the diabetes we thought it was. PNAS 103(33):12217–12118
32. Schwitzgebel VM (2014) Many faces of monogenic diabetes. J Diab Invest 5:121–133
33. Gao R, Liu Y, Gjesing AP, Hollensted M, Wan X, He S, Pedersen O, Yi X, Wang J, Hansen T (2014) Evaluation of a target region capture sequencing platform using monogenic diabetes as a study-model. BMC Genet 15:13. doi:10.1186/1471-2156-15-13
34. Ellard S, Lango AH, De Franco E, Flangan SE, Hysenaj G, Colclough K, Houghton JA, Shepherd M, Hattersley AT, Weeden MN, Caswell R (2013) Improved genetic testing for monogenic diabetes using targeted next-generation sequencing. Diabetologia 56:1958–1963
35. Haaland WC, Scaduto DI, Maldonado MR, Mansouri DL, Nalini R, Iyer D, Patel S, Guthikonda A, Hampf CS, Balasubramanyam A, Metzker ML (2009) A-β—subtype of ketosis-prone diabetes is not predominantly a monogenic diabetic syndrome. Diab Care 32:873–877
36. Bonnefond A, Durand E, Sand O, De Graeve F, Gallina S, Busiah K, Lobbens S, Simon A, Chantelot BC, Letourneau L, Scharfmann R, Delplanque J et al (2010) Molecular diagnosis of neonatal diabetes mellitus using next-generation sequence of the whole exome. PLoS ONE 5: e13630
37. Asha HS, Chapla A, Shetty S, Thomas N (2015) Next-generation sequencing-based genetic testing for familial partial lipodystrophy. AACE Clin Case Rep 1(1):e28–e31
38. Erlich HA, Valdes AM, McDevitt SL, Simen BB, Blake LA, McGowan KR, Todd JA, Rich SS, Noble JA, Type 1 Diabetes Genetics Consortium (T1DGC) (2013) Next generation sequencing reveals the association of DRB3*02:02 with type 1 diabetes. Diabetes 62:2618–2622
39. Lee HS, Briese T, Winkler C, Rewers M, Bonifacio E, Hyoty H, Pflueger M, Simell O, She JX, Hagopian W, Lernmark A et al (2013) Next-generation sequencing for viruses in children with rapid-onset type 1 diabetes. Diabetologia 56:1705–1711
40. Kramna L, Kalarova K, Oikarinen S, Purusiheimo JP, Ilonen J, Simelll O, Knip M, Veijola R, Hyoty H, O Cinek (2015) Gut virome sequencing in children with early islet autoimmunity. Diab Care 38:930–933

41. Talmud PJ, Cooper JA, Morris RW, Dudbridge F, Shah T, Engmann J, Dale C, White J, McLachlan S, Zabaneh D, Wong A, Ong KK, Gaunt T, Holmes MV, Lawlor DA et al (2015) Sixty-five common genetic variants and prediction of type 2 diabetes. Diabetes 64:1830–1840
42. Tanaka D, Nagashima K, Sasaki M, Funakoshi S, Kondo Y, Yasuda K, Koizumi A, Inagaki N (2013) Exome sequencing identifies a new candidate mutation for susceptibility to diabetes in a family with highly aggregated type 2 diabetes. Mol Genet Metab 109:112–117
43. Rung J, Cauchi S, Albrechtsen A, Shen L, Rocheleau G, Cavalcanti-Proenca C, Bacot F, Balkau B, Belisle A et al (2009) Genetic variant near IRS1 is associated with type 2 diabetes, insulin resistance and hyperinsulinemia. Nat Genet 41:1110–1115
44. Lyssenko V, Laakso M (2013) Genetic screening for the risk of type 2 diabetes. Diab Care 36: S120–S126
45. Kwak SH, Jang HC, Park KS (2012) Finding genetic risk factors of gestational diabetes. Genomics Inf 10:2390243
46. Bonnefond A, Philippe J, Durand E, Muller J, Saeed S, Arsian M, Martinez R, De Graeve F, Dhennin V, Rabearivelo I, Polak M, Cave H et al (2014) Highly sensitive diagnosis of 43 monogenic forms of diabetes or obesity through one-step PCR-based enrichment in combination with next-generation sequencing. Diab Care 37:460–467
47. Lieber DS, Vafai SB, Horton LC, Slate NG, Liu S, Borowsky ML, Calvo SE, Schmahmann JD, Mootha VK (2012) A typical case of Wolfram syndrome revealed through targeted exome sequencing in a patient with suspected mitochondrial disease. BMC Med Genet 13:3. doi:10.1186/1471-2350-13-3
48. Komers R, Xu B, Fu Y, McCelland A, Kantharidis P, Mittal A, Cohen HT, Cohen DM (2014) Transcriptome-based analysis of kidney gene expression changes associated with diabetes in OVE26 mice, in the presence and absence of losartan treatment. PLoS ONE 9(5):e96987. doi:10.1371/journal.pone.0096987
49. Kelly K, Liu Y, Zhang J, Goswami C, Lin H, Dominguez JH (2014) Comprehensive genomic profiling in diabetic nephropathy reveals the predominance of proinflammatory pathways. Physiol Genomics 45:710–719
50. Brennan EP, Morine MJ, Walsh DW, Roxburgh SA, Lindenmeyer MT, Brazil DP, Gaora PO, Roche HM, Sadlier DM, Cohen CD, GENIE Consortium, Godson C, Martin F (2012) Next-generation sequencing identified TGF-β1-associated gene expression profiles in renal epithelial cells reiterated in human diabetic nephropathy. Biochim Biophys Acta 1822:589–599
51. Pezzolesi MG, Krolewski AS (2013) The genetic risk of kidney disease in type 2 diabetes. Med Clin N Am 97:91–107
52. Kandpal RP, Rajasimha HK, Brooks MJ, Nellissery J, Wan J, Qian J, Kern TS, Swaroop A (2012) Transcriptome analysis using next generation sequencing reveals molecular signatures of diabetic retinopathy and efficacy of candidate drugs. Mol Vis 18:1123–1146
53. Sridhar GR, Lakshmi G (2015) Epigenetics and diabetes. In: Sridhar GR (ed) Advances in diabetes: Novel Insights. The Health Sciences Pub, N Delhi p 81–91
54. Wang J, Wu Z, Lif D, Li N, Dindot SV, Satterfield MC et al (2012) Nutrition, epigenetics and metabolic syndrome. Antioxidation Redox Signal 17:282–301
55. Ong FS, Lin JC, Das K, Grosu DS, Fan JB (2013) Translational utility of next-generation sequencing. Genomics 102:137–139
56. Salbaum JM, Kappen C (2011) Diabetic embryopathy: a role for the epigenome? Birth Defects Res A Clin Mol Teratol 91:770–780
57. Reddy MA, Natarajan R (2011) Epigenetic mechanisms in diabetic vascular complications. Cardiovasc Res 90:421–429
58. Latrelle M, Hausser J, Stutzer I, Zhang Q, Hastoy B, Gargani S et al (2014) MicroRna-7a regulates pancreatic β-cell function. J Clin Invest 124:2722–2735
59. Wren JD, Garner HR (2005) Data-mining analysis suggests an epigenetic pathogenesis for type 2 diabetes. J Biomed Biotechnol 2005(2):104–112

60. Miao P, Chen Z, Genuth S, Paterson A, Zhang L, Wu X et al (2014) Evaluating the role of epigenetic histone modifications in the metabolic memory of type 1 diabetes. Diabetes 63:1748–1762

61. Zhong X, Liao Y, Chen L, Liu G, Feng Y, Zeng T, Zhang J (2015) The microRNs in the pathogenesis of metabolic memory. Endocrinology 156(9):3157–3168. doi: 10.1210/en.2015-1063

62. Karlsson FH, Tremaroli V, Nookaew I, Bergstrom G, Behre CJ et al (2013) Gut metagenome in European women with normal impaired and diabetic glucose control. Nature 498:99–103

63. Sridhar GR (2015) Microbiota and metabolic syndrome. In: Bajaj S et al (eds) ESI handbook of endocrinology. Jaypee Pub, Delhi, pp 122–138

64. Sekirov I, Shannon L, Russell SL, Caetano MA, Finlay BB (2010) Gut microbiota in health and disease. Physiol Rev 90:859–904

65. Karlsson F, Tremaroli V, Nielsen J, Backhed F (2013) Assessing the human gut microbiota in metabolic diseases. Diabetes 62:3341–3349

66. Kim BS, Jeon YS, Chun J (2013) Current status and future promise of the human microbiome. Ped Gastreoenterol Hepatol Nutr 16:71–79

67. Teutsch SM, Bradley LA, Palomaki GE, Haddow JE, Piper M, Calonge N, Dotson WD, Douglas MP, Berg AO (2009) The evaluation of genomic applications in practice and prevention (EGAPP) initiative: methods of the EGAPP working group. Genet Med 11:3–14

68. Tang Y, Axelsson AS, Spegel P, Andersson LE, Mulder H, Groop LC, Renstrom E, Rosengren AH (2014) Genotype-based treatment of type 2 diabetes with an alpha2α-adrenergic receptor antagonist. Science Transl Med 6:257ra139

69. Jyothi KS, Srinivas K, Sridhar GR, Rao BS, Apparao A (2010) Plant insulin: an in silico approach. Intl J Diab Dev Countries 30:191–193

70. Annadurai RS, Jayakumaar V, Mugasimangalam RC, Katta MA, Anand S, Gopinathan S, Sarma SP, Fernandes SJ, Mullapudi N, Murugesan S, Rao SN (2012) Next generation sequencing and de novo transcriptome analysis of Costus pictus D. Don, a non-model plant with potent anti-diabetic properties. BMC Genom 13:663. doi:10.1186/1471-2164-13-663

71. Tang ZH, Fang Z, Zhou L (2013) Human genetics of diabetic vascular complications. J Genet 92(3):677–694

72. Jameson JL, Longo DL (2015) Precision medicine-personalized, problematic, and promising. N Engl J Med 372:2229–2234

73. Fall T, Xie W, Poon W, Yaghootkar H, Magi R, The GENESIS Consortium, Knowles JW, Lyssenko V, Weedon et al (2015) Using genetic variants to assess the relationship between circulating lipids and type 2 diabetes. Diabetes 64:2676–2684

74. Swerdlow DL, Sattar N (2015) Blood lipids and type 2 diabetes risk: can genetics help untangle the web? Diabetes 2015(64):2344–2345

75. Phimister EC (2015) Curating the way to better determinants of genetic risk. N Engl J Med 372:2227–2228

76. Rehm HL, Berg JS, Brooks LD, Bustamante CD, Evans JP, Landrum MJ, Ledbetter DH, Maglott DR, Martin CL et al (2015) ClinGen-the clinical genome resource. N Engl J Med 372:2235–2242

77. Kohane IS (2015) Ten things we have to do to achieve precision medicine. Science 349:37–38

78. Bielinski SJ, Pathak J, Weinshilboum RM, Wang L, Lyke KJ, Ryu E, Targonski PV, Van Norstrand MD, Hathcock MA, Takahashi PY, McCormick JB, Johnson KJ et al (2014) Preemptive genotyping for personalized medicine: design of the right drug, right dose, right time-using genomic data to individualise treatment protocol. Mayo Clin Proc 89:25–33

79. Editorial (2015) Data overprotection. Nature 522:391–392

80. Sarewitz D (2015) Science can't solve it. Nature 522:413–414

81. Veltman JA, Lupski JR (2015) From genes to genomes in the clinic. Genome Med 7:78

82. Vincent AT, Charette S (2015) Who qualifies to be a bioinformatician? Front Genet 6:164. doi:10.3389/fgene.2015.00164

83. Smith DR (2015) Broadening the definition of a bioinformatician. Front Genet 6:258. doi:10.3389/fgene.2015.00258

84. Middha S, Lindor NM, McDonnell SK et al (2015) How well do whole exome sequencing results correlate with medical findings? A study of 89 Mayo Clinic Biobank samples. Front Genet 6:244. doi:10.3389/fgene.2015.00244

Chapter 2
Coralyne Targets Proteases Involved in Cancer Progression: An In Silico Study

Seema Kumari, Anil Badana, V. Gayatridevi, P. Nagaseshu, M. Varalakshmi and Rama Rao Malla

Abstract Molecular docking has a significant application in finding the targets involved in cancer meta stasis. A positive correlation between the vigourness of tumor and expression of various proteases has been established, which includes serine proteases like furin and uPA, matrix metalloproteinases such as membrane type 1 MMP and tissue inhibitor of metalloproteinases, cysteine proteases such as cathepsin B & S and aspartate protease like cathepsin D. Virtual screening based on structure and post-screening analysis are routinely used in search of novel lead compound and its optimization. In the present study, the binding energy of coralyne with various metastatic proteases was analyzed using in silico docking tools such as iGEMDOCK v2.1, hex v6.3 and patch dock. The analysis of results indicates that coralyne exhibited significantly good binding affinity with furin and uPA predicting the possibility of coralyne in regulating cancer invasion and metastasis. Further, protein-protein network was analyzed using STRING version 10 based on KEGG pathways and clustered into groups using on MCL and k-mean algorithms to unfold its interacting partners proteins in cancer metastasis.

Keywords Coralyne · Metastatic proteases · iGEMDOCK · Hex · STRING

2.1 Introduction

Cancer is an unchecked growth of cells, which turns into serious disease conditions. Thus, development of a new drug against cancer is a vital step. Proteolysis is one of the core biological event involved in metastasis, cell proliferation, angiogenesis and apoptosis. Serine proteases, the largest human protease gene family, mediate a multifariousness events relevant to metastasis. Cancer invasion and metastasis are a result of degradation of basement membrane mainly by metallo-

S. Kumari · A. Badana · V. Gayatridevi · P. Nagaseshu · M. Varalakshmi · R.R. Malla (✉)
Department of Biochemistry, GIS, GITAM University, Visakhapatnam, India
e-mail: dr.rrmalla@gmail.com

© The Author(s) 2016
P.V. Lakshmi et al. (eds.), *Computational Intelligence Techniques in Health Care*, Springer Briefs in Forensic and Medical Bioinformatics,
DOI 10.1007/978-981-10-0308-0_2

19

proteinases (MMPs) [1]. Mammalian cysteine proteases are confined to lysosome such as cathepsins B, S, H, and L or in the cytosol like calpains. Reports have suggested that there is an interrelationship between the activity of cysteine proteases and aggressiveness of cancer as they can degrade both intracellular and extracellular matrix (ECM) proteins. Cathepsin B is involved in the disintegration of connective tissue and basement membrane which lead to cancer metastasis [2]. Aspartic proteases are enzymes which contain two lobes bridged by a cleft containing the catalytic site with two aspartate residues. Cathepsin-D is a pervasive aspartic endoprotease distributed in lysosomes. Studies suggest that variation in expression of cath-D may play an important role in cancer metastasis by favoring the growth of micro-metastases [3]. Furin is an important member of the family of pro-protein processing enzyme and highly expressed in a variety of tumors. Many proteins which are closely related to tumor development, including Notch, Wnt, MT1-MMP and VEGF, etc., are processed by furin. Thus, furin expression can be used as the marker of tumor progression or as prognostic indicators [4]. Alkaloids are secondary metabolites with the wide range of pharmacological properties. Coralyne [($C_{22}H_{22}O_4N^+$) 5, 6, 7, 8, 13, 13a hexadehydro-8-methyl-2, 3, 10, 11-tetramethoxy berberinium] is a crescent-shaped, planar heterocyclic isoquinoline alkaloid with broad biological activities. It is reported to have anti-leukemic activity and relatively low toxicity [5]. Cancer bioinformatics is a field which provides classification, clustering algorithmic and soft computing techniques to understand and predict possible early markers of cancers. Bio-computation is an important approach in drug designing as it speed up the process of drug designing and provide the platform to identify novel lead compounds. Docking of the lead molecule with the receptor at the possible drug action complementarity sites are responsible for pharmaceutical effect [6]. In the present study, the binding ability of coralyne with various metastatic proteases was analyzed using in silico docking. Further, a protein-protein network was analyzed using STRING version 10 based on KEGG pathways and clustered into groups using MCL and k-mean algorithms to unfold Furin interaction proteins in cancer metastasis.

2.2 Materials and Method

2.2.1 Retrieval of Receptors and Designing of Ligand

The crystal structure of the receptors was retrieved from Protein Data Bank (PDB). The PDB IDs of serine proteases: Furin (1P8J), uPA (4MNW), matrix metalloproteinases: MT1-MMP (1BQQ), TIMP-1 (1V96), TIMP-2 (1BR9), MMP-2 (1CK7), MMP-9 (1L6J), cysteine proteases: Cathepsin-B (2IPP), Cathepsin-S (4P6E), aspartate proteases: Cathepsin-D (1LYB). The 2D structure of coralyne and berberine was designed using Chemi-informatic software and converted to 3D structures.

2.2.2   Molecular Docking with iGEMDOCK v2.1

The proteases were docked with the ligands under docking accuracy settings (GA parameters) with binding site radius 8° A (X = 8.3, Y = 8.3 and Z = 8.3)° A each; population size 200; solutions 3; generation 70. The hydrophobic and electrostatic preference were set to 1.00. The empirical scoring function of iGEMDOCK was determined at: Fitness = van der Waal energy (vdW) + hydrogen bonding energy (Hbond) + electro statistic energy (Elec).

2.2.3   Docking with Hex v6.3

Hex docking of proteases and ligand was performed using the method of [7]. Docking was done by bringing the ligand into the vicinity of the receptor. Based on the energy minimization and 3D shape optimization, the Hex v6.3 docking control was set as follows: Correlation Type: Shape only, FFT Mode: 3D, Grid Dimension: 0.6, Solutions: 2000, Step Size: 7.5, Receptor Range: 180, Ligand Range: 180, Step Size: 7.5, Step Size: 5.5, Twist Range: 360, Scan Step: 0.8, Distance Range: 40, Sub Steps: 0.

2.2.4   Patchdock

Patchdock server (www.bioinfo3d.cs.tau.ac.il/PatchDock) was used to compute the scores of the docked complexes. The server depends on the principle of surface patch, molecular shape matching, filtering and scoring. 3D structures of proteases and ligand were submitted in PDB format with a protein-ligand parameter in Patchdock as the given input for the molecular docking. The output obtained is in the order of highly shape complementarity criteria which is given in the form of the score. The Patchdock results were evaluated for active site interactions by UCSF-chimera software [6].

2.2.5   STRING

STRING (*Search Tool for the Retrieval of Interacting Genes/Proteins*) identifies protein-protein interaction partners. This tool offers pre-computed interaction data derived from varied sources, such as high thought-put, experimental and literature data and computational predictions. The prediction methods selected for the neighborhood, gene fusion, co-occurrence, co-expression, experiments, database and text mining. Allows grouping of interacting molecules into clusters using MCL

(Markov clustering) and k-mean algorithms in the advanced mode. Therefore, this tool was used to query, retrieve and analyze the furin protein interaction network with the interactions restricted to those available for *Homosapiens*.

2.3 Results

In the present study, iGEMDOCK v2.1, hex v6.3 and patchdock were used to dock coralyne a lead target and berberine as standard with serine, cysteine, aspartic acid proteases and matrix metalloproteinases using molecular docking analysis.

2.3.1 *Effect of Coralyne on Serine Proteases*

Furin is a serine protease expressed by all tissues and cells. It is a membrane bound, calcium-dependent endoprotease. It plays a crucial role in the physiological function of embryogenesis, homeostasis and even in diseases such as cancer. Thus, finding potential inhibitors of furin may provide a promising approach in the treatment of cancer metastasis [8]. Urokinase-type plasminogen activator (uPA), which is involved in the conversion of plasminogen to plasmin, which directly or indirectly dissociates the extracellular matrix (ECM) via activation of (pro-MMPs), thus involved in cancer metastasis [9].

In the present study, docking of serine proteases like furin and uPA with coralyne was carried out by iGEMDOCK v2.1, hex v6.3 and patchdock (Figs. 2.1, 2.2, 2.3). The results obtained for total binding energy using iGEMDOCK was −96.95 and −91.80 kcal/mol for furin and uPA, respectively. Active site amino acids associated were: Thr 226 (−2.6), Pro 170 (−6.1), Gly 208 (−5.6), Asn 209 (−12), Asn 209 (−5.3), Ser 210 (−7.6), Gly 211 (−8); Gln 154 (−3.5), Phe 21 (−4.3), Thr 22 (−6.3), Thr 23 (−5.4), Arg 72 (−10.2) Thr 77 (−4), Gln 154 (−13.1). Berberine which was used as standard protoberberines alkaloid exhibited binding energy of −86.27 with active site amino acids as Arg 207 (−3.3), Pro 170 (−4.7), Ile 205 (−4.5), Ile 205 (−8.6) Arg 207 (−15.9), Ser 210 (−4) with furin and total binding energy of −81.39 with active site amino acids as Gly 69 (−3.5), Arg 70 (−5.1), Arg 70 (−4), Arg 70 (−4), Ser 71 (−4.2), Arg 72 (−4.7), Leu 73 (−7.4), Glu 153 (6.1), Gln 154 (−5.6), Leu 155 (−10.1) with uPA. Coralyne exhibited high affinity for furin and uPA, which may be due to presence of Thr and Gln at active site. Docking of aforesaid serine proteases using hex v6.3 reveals the binding energy as −290.06 and −281.26 for coralyne and berberine −272.19 and −234.54 kcal/mol (Table 2.1). Docking with patchdock showed score of 5784 and

Table 2.1 Docking score (kcal/mol) of coralyne and berberine with proteases

Proteases		iGEMDOCK v2.1						Hex 6.3	
		Total binding energy		Vander waal force		Hydrogen bond		E-value	
		Coralyne	Berberine	Coralyne	Berberine	Coralyne	Berberine	Coralyne	Berberine
Serine protease	Furin (1P8J)	−96.95	−86.27	−83.69	−87.26	−13.26	–	−290.06	−272.19
	uPA (4MNW)	−91.80	−81.39	−81.21	−67.75	−10.62	−13.64	−281.26	−234.54
Matrix metallo proteinase	MT1-MMP (1BQQ)	−64.59	−72.89	−60.17	−67.82	−4.42	−5.07	−261.17	−271.29
	TIMP-1 (1V96)	−87.5	−92.23	−84.56	−87.23	−2.94	−5.00	−251.63	−262.13
	TIMP-2 (1BR9)	−74.52	−74.98	−61.61	−63.94	−12.91	−11.04	−259.09	−263.89
	MMP-2 (1CK7)	−83.68	−94.55	−76.79	−93.16	−6.83	−1.38	−257.21	−292.92
	MMP-9 (1L6J)	−86.76	−93.30	−82.25	−90.70	−4.51	−2.60	−262.59	−297.67
Cysteine proteases	Cathepsin-B (2IPP)	−76.47	−71.96	−76.47	−65.86	–	−6.10	−269.15	−253.28
	Cathepsin-S (4P6E)	−67.61	−65.77	−64.11	−58.77	−3.50	−7.00	−218.62	−221.09
Aspartate proteases	Cathepsin-D (1LYB)	−69.83	−69.53	−65.21	−67.03	−4.61	−2.50	−208.32	−227.63

4796 for furin and uPA with coralyne; 5222 and 5177 with berberine. These results indicate that among above mentioned serine proteases furin has higher affinity for coralyne in comparison with berberine compared to uPA protease.

2.3.2 Effect of Coralyne on Matrix Metalloproteinases

Metalloproteinases (MMPs) are involved in cancer invasion and metastasis by degrading basement membrane which is mainly carried out by Membrane-bound MMPs (MT1) degrade ECM macromolecules, such as collagen I and III, laminin, vitronectin, fibronectin and proteoglycans [10, 11]. The major endogenous regulators of MMP activities is tissue inhibitors of metalloproteinases (TIMPs). Studies on the expression of MMP and TIMP have shown that cancer progression is well associated with the expression and/or overexpression of certain MMPs and TIMP-1 [12]. Figure 2.1 demonstrate the binding energy of matrix metalloproteinases MT1-MMP (1BQQ), TIMP-1 (1V96), TIMP-2 (1BR9), MMP-2 (1CK7), MMP-9 (1L6J) with coralyne using iGEMDOCK was revealed as −64.59 (Ser 1004 (−3.3),

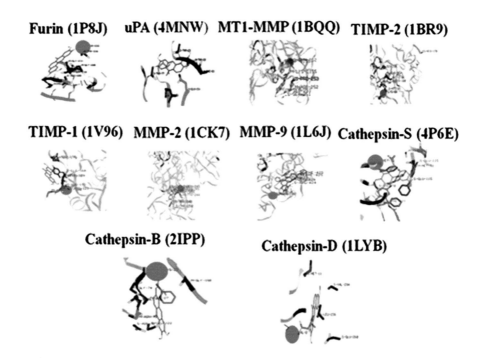

Fig. 2.1 Docking poses of proteases with coralyne using iGEMDOCK v2.1

Trp 1151 (−6.2), Asn 229 (−5.2), Tyr 261 (−6.7), Pro 1005 (−7.6), Pro 1005 (−8.2), Arg 1132 (10.9)); −87.5 (His 178 (−2.9), Lys 88 (−8.1), Ile 96 (−5.6), Ser 100 (−6.8), Ser 100 (−5.9)), −74.52 (Gln 10 (−3.4), Asn 14 (−6.9), Ser 141 (−2.5), Gln 10 (−8.5), Asn 14 (−4.5), Lys 129 (−12.1), Ile 130 (−5.5), Thr 131 (−5.4), Glu 145 (−9.8)), −83.68 (Lys 187 (−2.6), Asp 370 (−4.2), Gly 371 (−8.1), Met 373 (−7.2), Phe 389 (−6.9), Asp 392 (−11.4) and −86.76 (Tyr 52 (−2.5), Asn 38 (−7), Leu 44 (−7.6), Tyr 48 (−6.8), Arg 51 (−4.1), Tyr 52 (−5), Arg 95 (−12.1), Gly 186 (−11.4)) which was higher than berberine −72.89 (Ser 189 (−3.5), Asp 1034 (−3.1), Tyr 203 (10.6), Lys 1041 (−12.5), Phe 1067 (−5.7)), −92.23 (Cys 70 (−2.5), Ser 179 (−2.5), Thr 97 (−9.9), Cys 99 (−6.4), Ser 100 (−5.4), Asp 174 (−6.3), His 178 (−9.9)), −74.98 (Ser 7 (−3.5), Ala 15 (−2.5), Arg 102 (−2.5), Cys 3 (−4.9), Ser 4 (−5.2), Cys 100 (−5.1), −94.55 (Pro 417 (−8.8), Ala 422 (−6.6), Ile 424 (−6.5), Thr 426 (−4.7), Leu 508 (−5.9)) and −93.30 (Arg 51 (−2.7), Tyr 48 (−4.2), Tyr 52 (−4.1), Met 94 (−8.5), Pro 97 (−4.3), Gly 186 (−6)) Table 2.1. Further, results obtained from docking using hex v6.3 reveals binding energies with coralyne (−261.17, −251.63, −259.09, −257.21 and −262.59); berberine (−271.29, −262.13, −263.89 and −292.92), respectively with aforesaid matrix metallopro-teinases (Table 2.1) and Fig. 2.2. Patch dock showed score of 4446, 5148, 4330, 4422 and 5162 with coralyne and 3446, 5253, 5138, 3658 and 4556 with berberine as shown in Table 2.2 and Fig. 2.3. Coralyne exhibited high binding affinity with MMPs than berberine.

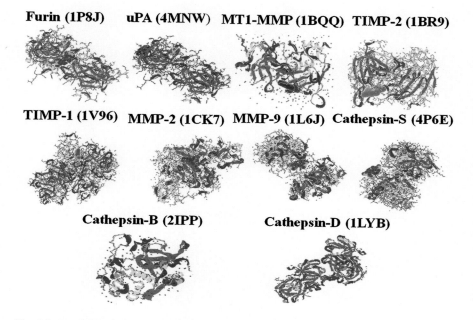

Fig. 2.2 Docking poses of proteases with coralyne using Hex v6.3

Table 2.2 Docking score of coralyne and berberine with proteases using patchdock

Proteases		Patchdock					
		Score		Area		Atomic contact energy (ACE)	
		Coralyne	Berberine	Coralyne	Berberine	Coralyne	Berberine
Serine protease	Furin (1P8J)	5278	−82.01	679.60	−76.82	−178.23	−5.07
	uPA (4MNW)	4422	−65.77	565.80	−289.48	−289.48	−7.00
Matrix metallo proteinase	MT1-MMP (1BQQ)	4446	−81.39	544.60	−67.75	−256.46	−13.64
	TIMP-1 (1V96)	4848	595.30	595.30	−87.23	−273.69	−5.00
	TIMP-2 (1BR9)	4330	−74.98	563.80	−63.94	−455.57	−11.04
	MMP-2 (1CK7)	5222	−94.55	687.70	−93.16	332.28	−1.38
Cysteine proteases	MMP-9 (1L6J)	4662	−93.30	655.60	−90.70	−411.35	−2.60
	Cathepsin-B (2IPP)	4318	−71.96	637.00	−65.86	−352.77	−6.10
	Cathepsin-S (4P6E)	4446	−86.27	544.60	−87.26	−242.61	–
Aspartate proteases	Cathepsin-D (1LYB)	5112	−69.53	670.00	−67.03	242.61	−2.50

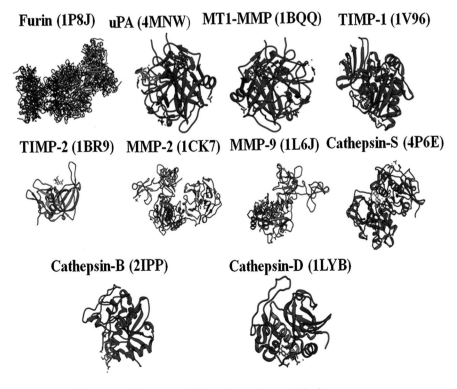

Fig. 2.3 Docking poses of proteases with coralyne using patchdock

2.3.3 Effect of Coralyne on Cysteine and Aspartate Proteases

Cysteine proteases are a proteolytic enzymes which are characterized by the presence of cysteine residue at an active site. They are confined to lysosome and cytosol. Reports suggested that there is a correlation between the activity of cysteine proteases and cancer metastasis as these proteases are involved in dissociation of extracellular matrix proteins [13]. Cathepsin B involved in degradation and reformation of basement membrane which may lead to cancer growth, invasion and metastasis [3]. Aspartic proteases are enzymes with two lobes bridged by a cleft containing with two aspartate residues at the catalytic site. Cathepsin-D is one of the aspartic endoprotease which is ubiquitous in distribution and confined to lysosomes. Studies suggests that variation in the expression levels of this protease may be responsible in cancer metastasis [14]. In the present study, cathepsin-B (2IPP) and cathepsin-S (4P6E) were considered for cysteine proteases and cathepsin-D (1LYB) as aspartate proteases. Figure 2.1 demonstrate the binding efficiency of coralyne with cathepsin B and S as −76.47 (Trp 30 (−5.1), Asn 72 (−4.9), Gly 73 (−9.8), His 199 (−9.8), Ala 200 (−6)), −67.61 (Lys 236 (−3.5), Ser 80 (−9.8), Ser 80(−5.3), Gln 258 (−7.6), Glu 260 (−8.7), Val 3 (−4.7), Met 307 (−7.7)) and with berberine as −71.96 (Trp 26 (−3.5), Gly 165 (−3.5), Gly 165 (−3.5), Ser (213 (−2.5)), −65.77 (Ser 7 (−3.5), Ala 15 (−2.5), Phe 70 (−4.6), Met 71 (−7.4)) with iGEMDOCK and −290.06 and −253.28 with coralyne; −253.28 and −221.09 with berberine (Table 2.1) using hex v6.3 Fig. 2.2 and score of 4318 and 4446 for coralyne; 3970 and 5440 with berberine using patchdock (Table 2.2) and Fig. 2.3. Aspartic acid proteases exhibited −69.83 (Leu 67 (−2.5), Ser 89 (−4.7), Gln 98 (−3.5), Glu 260 (−6.4), Met 72 (−3.5), Val 294 (−7.6), Leu 236 (−2.5)) for coralyne; −69.53 (Cys 25 (−3.8), Trp 26 (−3.5), Gly 165 (−3.5), Ser (213 (−2.5), Gly 69 (−7.1), Phe 70 (−4.6), Met 71 (−7.4), Val 258 (−5.3), Leu 236 (−2.5)), for berberine using iGEMDOCK as shown in Fig. 2.1. Docking with hex Fig. 2.2 reveals coralyne exhibited binding affinity as −208.32 and −227.63 for berberine (Table 2.1) and score of 5112 and 4700 for coralyne and berberine (Table 2.2) using patchdock Fig. 2.3. More the negative value higher is the affinity, thus among all the proteases serine proteases exhibited high binding energy and as the binding affinity of coralyne was higher than berberine it can be used as antimetastatic alkaloid targeting proteases.

2.3.4 Identification of Furin Interaction Protein Partners

As furin is having high binding affinity with coralyne, its interaction partners were identified using STRING. The results showed that furin exhibited interaction with Notch, TIMP, MMPs and VEGF which are involved in regulation development of cell. Further, protein-protein interacting clusters were analyzed using K-mean [15].

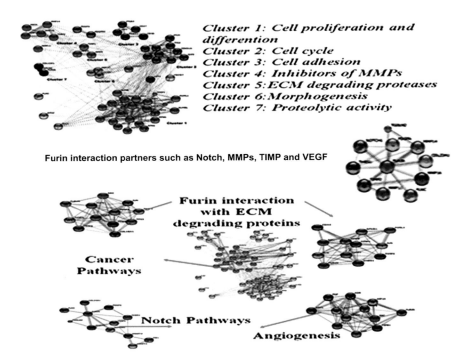

Fig. 2.4 Furin clustering using K-mean and interaction with metastatic proteins using KEGG pathway

The results showed that Cluster 1 is involved in cell proliferation and differentiation, cluster 2 regulates cell cycle, cluster 3 is involved in cell adhesion, cluster 4 are the inhibitors of MMPs, cluster 5 are ECM degrading proteases, cluster 6 is involved in morphogenesis and cluster 7 are involved proteolysis (Fig. 2.4).

2.4 Discussion

In the present study, it was analyzed that coralyne (ligand) exhibited high negative binding energy with furin, a serine protease, this indicates the high binding affinity between coralyne and furin. Reports suggest that furin which is an important member of the family of pro-protein processing enzymes and is highly expressed in various tumors. Thus, protein-protein network was analyzed to unfold its interaction proteins in cancer metastasis using STRING, which offers pre-computed interaction data derived from varied sources, such as, high thought-put, experimental data, literature data and computational predictions. It was analyzed that furin exhibited direct interaction with proteins involved in tumor development, invasion and metastasis including Notch, TIMP, MMPs, VEGF. Thus, furin can be an important target for controlling cancer metastasis. Clustering of various interacting proteins with furin was performed by k-mean algorithms. The clusters were characterized as proteins involved in cell development, differentiation, cell cycle, invasion and metastasis. Further, using KEGG pathway proteins involved in cancer pathway where furin showed direct or indirect interactions were predicted.

2.5   Conclusion

A significant interaction of coralyne with furin, a serine protease was observed compared to other proteases. Further, Notch, TIMP, MMPs and VEGF which are involved in cancer progression was predicted as protein interaction partners of furin. Thus, this study may be useful for evaluating the effect of coralyne on various mechanisms of different cancers, both in vitro and in vivo models and also to identify furin as the drug target.

Acknowledgements   The present research work was supported by UGC, (New Delhi) with file number (No.F.15-1/2013-2014/PDFWM-2013-2014-GE-AND-12376 (SA-II)) for Post-Doctoral Fellow for Woman. I would like to thanks, GITAM University for providing lab facilities.

References

1. Kumari S et al (2013) Anti-proliferative and metastatic protease inhibitory activities of protoberberines: an in silico and in vitro approaches. Process Biochem 48:1565–1571
2. Yang Y, Hong H et al (2009) Molecular imaging of proteases in cancer. Cancer Growth Metastasis 2:13–27
3. Rakashanda S et al (2012) Role of proteases in cancer: a review. Biotechnol Mol Biol Rev 7 (4):90–101
4. Ma YC et al (2014) Effect of furin inhibitor on lung adenocarcinoma cell growth and metastasis. Cancer Cell Int 14:43
5. Mathew AJ et al (2009) Docking studies on anticancer drugs for breast cancer using hex. In: Proceedings of the international multiconference of engineers and computer scientists, vol 1, pp 18–20
6. Islam MM et al (2009) RNA targeting through binding of small molecules: studies on t-RNA binding by the cytotoxic protoberberine alkaloid coralyne. Mol BioSyst 5:244–254
7. Ritchie D (2010) Hex 6.3 user manual, Laboratoirelorrain de recherché en informatique etses application, France, p 1
8. Fittler H et al (2015) Engineering a constrained peptidic scaffold towards potent and selective furin inhibitors. Chembiochem. (Epub ahead of print)
9. Tang L, Han X (2013) The urokinase plasminogen activator system in breast cancer invasion and metastasis. Biomed Pharmacother 67(2):179–182
10. Yana I, Seiki M (2002) MT-MMPs play pivotal roles in cancer dissemination. Clin Exp Metastasis 19:209–215
11. Bae MJ et al (2015) Evaluation of effective MMP inhibitors from eight different brown algae in human fibrosarcoma HT1080 cells. Prev Nutr Food Sci 20(3):153–161

12. Kousidou OCH et al (2004) Expression of MMPs and TIMPs genes in human breast cancer epithelial cells depends on cell culture conditions and is associated with their invasive potential. Anticancer Res 24:4025–4030
13. Gocheva V, Joyce JA (2007) Cysteine cathepsins and the cutting edge of cancer invasion. Cell Cycle 1:60–64
14. Berchem GJ et al (2002) Cathepsin-D affects multiple steps of tumor progression: proliferation, angiogenesis and apoptosis. Oncogene 51:5951–5955
15. Yin Y et al (2014) Bioinformatic analysis of miRNA expression patterns in TFF2 knock-out mice. Genet Mol Res 13(4):8502–8510

Chapter 3
Computational Intelligence Approach for Prediction of Breast Cancer using Particle Swarm Optimization: A Comparative Study of the Results with Reduced Set of Attributes

Kalagotla Satishkumar, T. Sita Mahalakshmi and Vedavathi Katneni

Abstract Data mining is termed as mining of relevant information from abundant volumes of data in order to promote the business activity and decision making capability. The exploratory data analysis is a similar technique for summarizing and identifying the patterns in the data. This exploratory data analysis is a statistical model like a regression model, discriminant model or a clustering model, which is built from the data and is utilised for prediction, classification, or hypothesis verification. We presented a rule discovery algorithm based on swarm intelligence in order to identify the standard production rules and apply the rule pruning mechanism to shorten the rule. The proposed algorithm uses Wisconsin Breast Cancer and Mammographic Mass Data. This algorithm when applied on Wisconsin Breast Cancer found to be more accurate compared to the existing classification algorithms namely C4.5 and Classification using Regression Trees. A comparative study has been performed with dimensionality reduction using entropy based Information gain measure.

Keywords Data mining · Classification · Association rules · Dimension reduction · Swarm intelligence · Particle swarm optimization

3.1 Introduction

Today the world is filled with tera bytes of data per every quarter hour. This is due to the advances in the field of data collection and data storage equipment. This abundant data is useful only when it is well organized and patterns are discovered from it. To do so, data mining techniques were developed to allow users to navigate through the data. Data mining has drawn its attention on machine learning. Data

K. Satishkumar (✉) · T. Sita Mahalakshmi · V. Katneni
GITAM University, Visakhapatnam, Andhra Pradesh, India
e-mail: satish7433@gmail.com

© The Author(s) 2016
P.V. Lakshmi et al. (eds.), *Computational Intelligence Techniques in Health Care*, Springer Briefs in Forensic and Medical Bioinformatics, DOI 10.1007/978-981-10-0308-0_3

mining is concerned with extracting valid patterns, where as machine learning is concerned with improving the performance of intelligent systems. One of the goals of data mining is prediction and description. Prediction makes use of available data and predicts the future values of interest where description finds out the pattern from the available data that describes it, and the subsequent presentation for user interpretation. Associations and classifications are the techniques of data mining that achieve prediction and description. Exploratory data analysis (EDA) is used to analyse the data. To enlighten the user to easily understand the characteristics of data, EDA uses both graphical and quantitative techniques. Graphical techniques include box plot, histogram, multi-variate chart, run chart, pareto chart, scatter plot etc. In contrast to graphical techniques the quantitative techniques used to describe the data are Median polish, Trimean and Ordination.

For various data mining tasks relevant data mining techniques have been witnessed. Among these, classification and association techniques are extensively studied. Classification task can be described as supervised learning and unsupervised learning. In Supervised classification, the classification is performed with samples of known class labels to generate a classifier. In Unsupervised classification, the classification is performed with samples of unknown class labels by grouping them into clusters based on data relevance to discover the basic format in data. The result of classification can be represented in the form of rules. Taking this into consideration, we developed an algorithm based on swarm intelligence technique for the extraction of classification rules. The rules are represented in the form of IF p THEN q where p is antecedent and q is consequent representing the class label. Our proposed algorithm identifies rules which are pruned based on post pruning technique to achieve shortened rules with high generalization and less computational effort.

3.2   Related Work

Pan and Yang [11] presented a survey on knowledge transfer where training and feature data exists in different feature space or follow a different distribution. The authors identified the relationship between transfer learning and other related machine learning techniques such as domain adaptation, multitask learning and sample selection bias, covariate. Simon et al. [13] reviewed four association rule set summarization techniques and compared the rule performance, applicability and weaknesses. Tripoliti et al. [16] presented an automated diagnosis of diseases based on the improvement of random forests classification algorithm. Sousa et al. [15] proposed the use of Particle Swarm Optimization (PSO) for classification rule discovery. The performance of the algorithm is tested against genetic algorithm, J48 a java implementation of C4.5.

A particle swarm optimization based simultaneous learning framework for clustering and classification was proposed by Liu et al. [8]. It is a three step process which involves an automatic cluster algorithm in the first step. The second step finds optimal cluster centre by using a relation matrix established through Bayesian theorem to find the relationship between cluster centres and classes. The third step is the classification of test data. A novel PSO based classification algorithm by Nouaouria and Boukadoum [10] uses a new particle position update mechanism and a new way to handle mixed attribute data based on particle position interpretation. Chen et al. [2] proposed a rule pruning algorithm namely Probabilistic Classification Based on Association (PCBA), compatible for Scoring Based Association algorithms (SBA) for solving the problem of associative classification with class imbalance in order to improve predictive accuracy of associative classification.

Liu et al. [7] presented an efficient post processing method to prune redundant rules. The design of Fuzzy Particle Swarm Optimization (FPSO) method was presented by Permana and Hashim [12] to improve fuzzy system performance and the speed of convergence. Mangat [9] presented a Swarm Intelligent (SI) based techniques for rule mining in medical domain. The PSO, Ant Colony Optimization (ACO)/PSO, and ACO/PSO with new quality measure of fitness were studied and their accuracies were compared. The remnant of the paper is organised as follows. Section 3.3 contains the algorithm description for rule generation. Section 3.4 describes the data set. The computed results are discussed in Sect. 3.5. Finally, Sect. 3.6 gives the conclusion.

3.3 Algorithm Description for Rule Generation

An important feature of artificial intelligence is to design machine that performs various tasks which normally requires a human intelligence. Every task solved by artificial intelligence has own features that distinguish it uniquely. Some features that were common to most of the tasks which require human intelligence like converting the problem into solvable form, heuristic search and its correlated sub problems, converting complex problems into simple problems, and learning through induction from past examples.

Swarm Intelligence mainly deals with global optimization issues. It is inspired from the nature. SI is defined as self organised natural or an artificial system with decentralised control and collective behaviour. It is a population based technique, in which each individual move in search space by interacting with each other emerging towards global optimum value. Ant colonies, bird flocking and animal herding are some of the examples of natural systems of SI. Swarm prediction is used in forecasting problems. The SI algorithms are PSO, ACO, Artificial bee colony algorithms etc. In our work we used PSO to represent a solution to a problem in **n** dimensional space.

Particle Swarm Optimization is an efficient search and optimization technique. PSO is conceptually simple as it uses basic mathematical operators. PSO is a population based algorithm, Ho et al. [4] have described mathematically, the swarm population of size $N_{popsize}$ with the swarm or particle i where $i \in \{1, 2, 3, \ldots N_{popsize}\}$ is associated with a position vector $x_i \in \{x_1^i, x_2^i, x_3^i, \ldots x_Z^i\}$ where the number of decision parameters of an optimal problem is Z. The best previous position is given by p_{best} which is the position that gives best fitness value. The particle i which was found in the parameter space is represented as $p_i \in \{p_1^i, p_2^i, p_3^i, \ldots p_Z^i\}$ the best position that the neighbourhood particles have ever found is g_i, is the global best particle in the parameter space. It is represented as $g_i \in \{g_1^i, g_2^i, g_3^i, \ldots g_Z^i\}$. At each iteration step (k + 1) the position vector of the ith particle namely x_i is updated by velocity of the ith particle v_i defined as $(v_1^i, v_2^i, v_3^i, \ldots v_Z^i)$.

$$v_Z^i(k+1) = \chi\left(v_Z^i(k)\right) + c_1 r_1\left(p_Z^i - x_Z^i(k)\right) + c_2 r_2\left(g_Z^i - x_Z^i(k)\right) \tag{3.1}$$

$$v_Z^i(k+1) = \frac{v_Z^i(k+1) * v_Z^{max}}{v_Z^i(k+1)}; \left(if |v_Z^i(k+1)| > v_Z^{max}\right) \tag{3.2}$$

$$x_Z^i(k+1) = x_Z^i(k) + v_Z^i(k+1) \tag{3.3}$$

By Bratton and Kennedy [1] where the constant χ is presented as 0.72984 and two positive constants c_1 and c_2 are 2.05. The velocity of the particle is limited to v_Z^{max} in the Zth coordinate direction. An iterative process is executed for swarm until the stopping condition is met. The influence of particles own experience on the velocity is presented in Eq. 3.1. The particles velocity is modified depending upon the neighbourhood experience.

Rules are used to represent the learned knowledge in an understandable way to perform analysis. For better interpretation of the data, the classifier extracts the information from the data and represents the extracted information in the form of rules. Rules are interpreted as If condition then Decision label. Condition is the conjunction of terms such that each term is composed of triplet containing attribute values, operator and value. The shorthand notation of rules is $p \rightarrow q$ where p is conjunction of terms. If the data tuple satisfy the condition part then data tuple is assigned with corresponding class label q. In our work we generated classification rules for both class1 and class2 for the sample data taken from Wisconsin breast cancer dataset.

Rule pruning reduces the complexity and increases the speed of the classifier. Rule pruning are of two types namely top down and bottom up. In top down trimming of rules is performed from root node to leaf node, in bottom up trimming of rule is performed from leaf node to root node. In our work, top down rule pruning is performed. Initially all attributes are involved in calculating predictive accuracy of a rule. Trimming of rule is done if the predictive accuracy of the rule is not affected.

Dimension reduction can be defined as reducing the number of random variables under consideration. Dimensionality reduction is an important problem in machine learning. They are two types of dimensionality reduction namely feature extraction and feature selection. Feature extraction is the transformation of high dimensionality data into fewer dimensions. Feature selection is selecting the most relevant ones from the existing attributes. Filter and wrapper are two techniques available for feature selection. The advantage of feature selection is that we can build better and faster learning machines. In our work we performed feature selection using Entropy based information gain measure.

According to Han et al. [3] Feature selection measure is a heuristic process. Let D is the set of training samples with class labels, the information needed to distinguish a tuple in D is given by product of probability that the tuple in D belongs to class C_i is estimated as $|C_{iD}|/|D|$ and log base 2 of probability that the tuple in D belongs to class C_i. The log function to the base 2 is used because the information is encoded in bits.

p_i is the ratio of ith tuple in D belongs to class C_i to total number of samples in D p_i is calculated as $p_i = \frac{|C_{i,D}|}{|D|}$

$$Info(D) = -\sum_{i=1}^{m} p_i \log_2(p_i) \qquad (3.4)$$

The normal information needed to distinguish a tuple in D when partitioned on attribute A having v distinct values, i.e., j over v partitions, for each partition info is calculated and multiplied with total samples of partition to total samples in D.

$$Info_A(D) = \sum_{j=1}^{v} \frac{|D_j|}{|D|} * Info(D_j) \qquad (3.5)$$

Gain of attribute is calculated as normal information needed to distinguish the tuple in D minus information needed to distinguish the tuple in D when partitioned on A.

$$Gain(A) = Info(D) - Info_A(D) \qquad (3.6)$$

3.3.1 Fitness Function

In our work the fitness of the particle is calculated by using False Positives (FP), which is good indication of better fitness when FP is low.

$$\text{Fitness} = \frac{1}{1 + FP} \qquad (3.7)$$

Where, FP is the negative tuples that were incorrectly labelled as positives.

Algorithm

Input: Train and test data.
Notations: pbest $\rightarrow$ the previous best particle
            gbest $\rightarrow$ the global best particle
            population $\rightarrow$ size of swarm
Procedure:
Step 1: Refer to the data file.
Step 2: Generate initial population and velocity.

Step 3: Calculate the fitness for each particle using Eq. 3.7 and make it as initial pbest.

Step 4: Maximum fitness particle is chosen as gbest particle.

Step 5: Update particle position based on Eq. 3.3 and velocity based on Eqs. 3.1 and 3.2.

Step 6: Store the pbest and gbest values in previous pbest and gbest respectively.

Step 7: If the stopping condition is not satisfied then repeat step 8 to step 13 else go to step14.

Step 8: Calculate the fitness for each particle using Eq. 3.7 and make it as pbest.

Step 9: Maximum fitness particle is chosen as gbest particle.

Step 10: Compare the previous pbest with pbest and previous gbest with gbest.

Step 11: Update the best pbest and gbest.

Step 12: Update particle position based on Eq. 3.3 and velocity based on Eqs. 3.1 and 3.2

Step 13: Store the pbest and gbest values in previous pbest and gbest.

Step 14: Update gbest as best rules.

Step 15: Perform rule pruning.

Step 16: Apply pruned rules on test data.

Step 17: Calculate accuracy.

Step 18: Stop.

Stopping criteria:

- If the iterations are greater than the maximum iterations specified.

3.4   Dataset Description

For this work, we considered the original Wisconsin breast cancer dataset, Mammographic Mass Data. These were taken from UCI Machine Learning Repository donated by Lichman [6], to distinguish malignant from benign cases. There were 699 instances with nine attributes and one class label field in total, after removing noisy data in the dataset, there were 683 instances with 444 benign and 239 malignant were considered for the experiment. Mammography Mass Data is used to predict the severity of a Mammography mass lesion from BI-RADS attribute and patient's age. It contains five attributes namely BI-RADS, age, shape, margin, density and a class label contain either benign or malignant. The data samples considered for training are 747 out of which 387 belong to benign and 360 belong to malignant. Test data contain 83 samples out of which 43 belong to benign and 40 belong to malignant.

Wisconsin breast cancer data set has nine integer type attributes and a class label. In clump thickness the normal cells are grouped in mono layers, where as the cancerous cells are tend to be grouped in multi layers. Variation of uniformity of cell shape and size plays a crucial role in detecting cancerous cells. In Marginal adhesion the normal cells are bounded where as the cancerous cells lose their ability, hence loss of adhesion results in malignant. Epithelial cells which are extremely enlarged may be a malignant cell. Bare Nuclei which is not surrounded by cytoplasm is seen in benign tumours. Bland chromatin describes the uniform texture of the nucleus seen in benign cells; Chromatin of cancerous cells tends to be coarser. The normal nucleoli are the small structures seen in nucleus. In normal nucleoli the nucleus is very small if visible, in cancerous cell the nucleoli are more prominent. Mitosis is the process in which the cell divides and replicates. Pathologists can determine the grade of cancer by identifying mitoses count. Class label contains benign and malignant.

3.5 Results and Discussions

3.5.1 K-Fold Cross Validation

K fold cross validation is one of the best measure of performance classifier. The Data set is split into K equal folds with equal class intervals, in which kth fold is taken as test set and $(k - 1)$ folds as training set. The process is run for K trails every time each fold is taken as test and remaining as training set. In our work we performed 10-fold cross validation for information gain based dimension reduction data and actual dimension data. The data is split into equal class intervals for 10 folds, out of 670 samples 440 belong to benign and 230 belong to malignant. 13 samples were removed from 683 samples to avoid class imbalance in 10 fold validation. Similarly, Mammographic Mass data consist of 830 samples out of which 747 train samples and 83 test samples. The generated rules on Wisconsin Breast Cancer is presented in Table 3.1 and comparative study of results on Wisconsin Breast Cancer (WBC) and Mammographic Mass Data are presented in Table 3.2. Ten fold confusion matrix with reduced dimension data and dimensional Wisconsin Breast Cancer data are presented in Figs. 3.1 and 3.2. Similarly, Tenfold

Table 3.1 Generated rules of Wisconsin breast cancer data

	Class1	Class2
Reduced dimension rules	If A2 <= 2.32 & A3 <= 2.40 & A5 <= 4.17 & A6 <= 3.22 & A7 <= 8.71	If A2 > 5.10 & A3 <= 11.32 & A5 > 2.26 & A6 <= 10.71 & A7 > 4.63
Reduced dimension pruned rules	If A5 <= 4.17 & A6 <= 3.2 & A7 <= 8.71	If A3 <= 11.32 & A6 <= 10.71 & A7 > 4.63
All dimension rules	If A1 <= 6.66 & A2 <= 2.622 & A3 <= 6.48 & A4 <= 6.44 & A5 <= 6.65 & A6 <= 3.79 & A7 <= 3.11 & A8 <= 2.79 & A9 <= 1.18	If A1 > 6.01 & A2 <= 9.41 & A3 <= 10.27 & A4 <= 8.94 & A5 <= 12.04 & A6 > 9.06 & A7 <= 8.45 & A8 <= 11.026 & A9 <= 3.37
All dimension pruned rules	If A1 <= 6.66 & A3 <= 6.48 & A4 <= 6.44 & A5 <= 6.65 & A6 <= 3.79 & A8 <= 2.79	If A3 <= 10.27 & A5 <= 12.04 & A6 > 9.06 & A8 <= 11.02

Table 3.2 Comparison of Information gain based reduction and high dimensional Wisconsin Breast cancer and mammographic mass data

Fold no	Information gain based Dimension Reduction (WBC)	Accuracy (%) (WBC)	All Dimensions (WBC)	Accuracy (%) (WBC)	Information gain based Dimension Reduction (MMD)	Accuracy (%) (MMD)	All Dimensions (MMD)	Accuracy (%) (MMD)
1	5	97	9	95.5	3	83.1	5	85.5
2	5	94	9	95.5	3	83.1	5	81.9
3	5	94	9	97	3	83.1	5	83.1
4	5	95.5	9	97	3	85.5	5	85.5
5	5	91	9	97	3	83.1	5	79.5
6	4	92.5	9	98.5	3	83.1	5	81.9
7	4	97	9	98.5	3	79.5	5	84.3
8	5	94	9	94	3	83.1	5	83.1
9	5	91	9	97	3	83.1	5	83.1
10	5	97	9	97	3	83.1	5	85.5
Average	–	94.3	–	96.7	–	83	–	83.4

Fig. 3.1 Ten fold reduced
dimension confusion matrix
of Wisconsin breast cancer

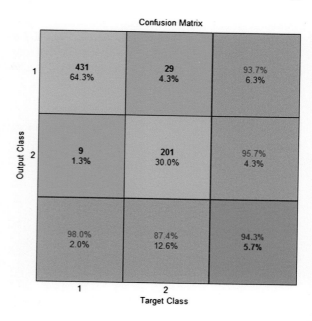

Fig. 3.2 Ten fold all
dimension confusion matrix
of Wisconsin breast cancer

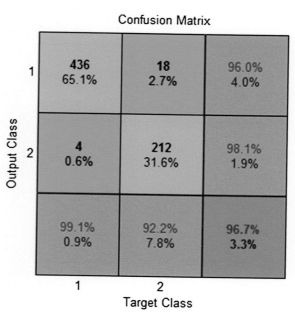

confusion matrix with reduced dimension data and dimensional data of
Mammographic Mass Data is presented in Figs. 3.3 and 3.4.

In our comparative study, we observed that our proposed algorithm when
applied on Wisconsin Breast Cancer gave 96.7 % accuracy. Mammographic Mass

Fig. 3.3 Ten fold reduced
dimension confusion matrix
mammographic mass data

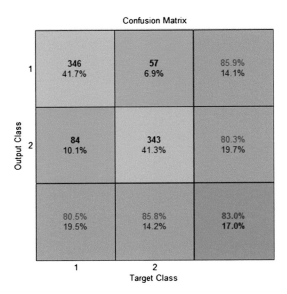

Fig. 3.4 Ten fold all
dimension confusion matrix
of mammographic mass data

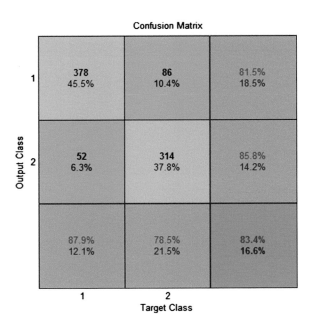

Data yeilded 83.4 % accuracy with all dimensions. With dimension reduction we achieved 94.3 % accuracy on Wisconsin Breast Cancer and 83 % accuracy on Mammographic Mass Data. From the above tabulated results it is clear that our PSO algorithm with all dimensions into consideration achieved good accuracy on Wisconsin Breast Cancer Data.

3.5.2 Performance Metrics

Performance metrics are the fundamental aspects in data mining to evaluate any technique. Some of the performance metrics for any classifier are Accuracy, Sensitivity, Specificity, Precision, Recall and F-measure. Accuracy is the percentage of correctly labeled tuples to the summation of correctly and incorrectly labeled tuples. Sensitivity is the percentage of truly labeled positives to the summation of positives labeled as positives and positives incorrectly labeled as negatives. Specificity is the percentage of true negatives to the summation of true negatives and false positives. Recall is same as sensitivity, precision is the percentage of correctly labelled positives to the total of positives and negatives labeled as positives. The performance metrics formulae and performance metrics calculations of Wisconsin Breast Cancer, Mammographic Mass data are presented in the Tables 3.3, 3.4. Comparison of average accuracies of Wisconsin Breast Cancer, Mammographic Mass data are presented in Table 3.5.

The value of F-measure for the proposed algorithms is nearer to 1 for all the classes of Wisconsin Breast Cancer, Mammographic Mass Data. which is the good indication for the performance of the classifier.

Table 3.3 Performance metrics formulae

Measure	Formula
Accuracy	$\dfrac{(TP + TN)}{(TP + TN + FP + FN)}$
Sensitivity	$\dfrac{(TP)}{(TP + FN)}$
Specificity	$\dfrac{(TN)}{(TN + FP)}$
Precision	$\dfrac{(TP)}{(TP + FP)}$
Recall	$\dfrac{(TP)}{(TP + FN)}$
F-measure	$F = \dfrac{(2 * PRECISION * RECALL)}{(PRECISION + RECALL)}$

Where
TP True positives which mean class1 labelled as class1
TN True negatives which mean class2 labelled as class2
FP False positive which mean class2 labelled as class1
FN False negative which mean class1 labelled as class2

Table 3.4 Performance metrics of Wisconsin breast cancer and mammographic mass data for both the classes

Performance metrics	Information gain based dimension reduction (WBC)		All dimensions (WBC)		Information gain based dimension reduction (MMD)		All dimensions (MMD)	
	Class1	Class2	Class1	Class2	Class1	Class2	Class1	Class2
Accuracy	0.9433	0.9433	0.9671	0.9671	0.8301	0.8301	0.8337	0.8337
Sensitivity	0.9370	0.9571	0.9603	0.9814	0.8586	0.8033	0.8147	0.8579
Specificity	0.9571	0.9370	0.9814	0.9603	0.8033	0.8586	0.8579	0.8147
Precision	0.9795	0.8739	0.9909	0.9217	0.8047	0.8575	0.8791	0.7850
Recall	0.9370	0.9571	0.9603	0.9814	0.8586	0.8033	0.8147	0.8579
F-measure	0.9578	0.9136	0.9753	0.9506	0.8307	0.8295	0.8456	0.8198

Table 3.5 Comparison of proposed algorithms Accuracy on different data sets

DataSet	PSO based on Information gain based dimension reduction (%)	PSO based on reduced dimensions (%)
Wisconsin breast cancer	94.3	96.7
Mammographic mass data	83	83.4

Table 3.6 Comparision of proposed methodology with other techniques

Authors	Technique	Accuracy (%)
Charoenchai Sirisomboonrat et al. [14]	C4.5	94.72
Lavanya and Rani [5]	CART	94.84

In Table 3.6 the proposed algorithm is compared with different techniques, when compared with other techniques our algorithm performance was good in terms of classification accuracy when applied on Wisconsin Breast Cancer which gave 96.7 % accuracy with PSO based on All dimension into consideration.

3.6   Conclusion

In this paper we presented a novel rule discovery algorithm based on PSO, and a comparative study was made with reduced set of attributes when applied the algorithm on Wisconsin Breast Cancer and Mammographic Mass Data. This proposed algorithm gave 96.7 % accuracy with all dimensions when experimented with ten fold cross validation on WBC. PSO algorithm based on all Dimensions into consideration gave best result on Wisconsin Breast Cancer dataset, Mammographic Mass Data. The best yield dimensionality algorithm namely PSO with All dimensions when applied on WBC gave 96.7 % accuracy which is compared with other rule extraction techniques like C4.5 and CART. Our algorithm performed far better when compared with these algorithms.

From the experimental results it is clearly observed that our proposed algorithm achieved 96.7 %accuracy on considering all dimension data, it means 96.7 %accurately our proposed algorithm detect the breast cancer and helps the patient to take the better treatment and the radiologist to provide effective treatment for patients. Our further work includes applying other techniques on the Breast cancer datasets to achieve better classification accuracy.

FUNDING

This work was supported by National Project Implementation Unit (NPIU) which provided research assistantship under Technical Education Quality Improvement Programme of Government of India (TEQIP-II).

Acknowledgement We are thankful to the review committee of Gitam University, whose comments allowed for a great deal of improvement of the work.

References

1. Bratton D, Kennedy J (2007) Defining a standard for particle swarm optimization. In: Proceedings of the IEEE Swarm Intelligence Symposium
2. Chen WC et al (2012) Increasing the effectiveness of associative classification in terms of class imbalance by using a novel pruning algorithm. Expert Syst Appl 39:12841–12850
3. Han J et al (2011) Data mining: concepts and techniques, san francisco. Morgan Kaufmann Publishers Inc., CA
4. Ho SL et al (2005) A particle swarm optimization-based method for multiobjective design optimizations. IEEE Trans Magn 41(5):1756–1759
5. Lavanya D, Rani DKU (2011) Analysis of feature selection with classification: breast cancer datasets. Indian J Comput Sci Eng (IJCSE)
6. Lichman M (2015) Breast cancer Wisconsin (Original) Data set, UCI machine learning repository. http://archive.ics.uci.edu/ml. Accessed 19 May 2015
7. Liu H et al (2011) A fast pruning redundant rule method using Galois connection. Appl Soft Comput 11:130–137
8. Liu R et al (2014) A particle swarm optimization based simultaneous learning framework for clustering and classification. Pattern Recog 47:2143–2152
9. Mangat V (2010) Swarm intelligence based technique for rule mining in the medical domain. Int J Comput Appl 4(1):19–24
10. Nouaouria N, Boukadoum M (2014) Improved global-best particles warm optimization algorithm with mixed-attribute data classification capability. Appl Soft Comput 21:554–567
11. Pan SJ, Yang Q (2010) A survey on transfer learning. IEEE Trans Knowl Data Eng 22 (10):1345–1359
12. Permana KE, Hashim SZM (2010) Fuzzy membership function generation using particle swarm optimization. Int J Open Probl Comput Sci Math 3(1):27–41
13. Simon GJ et al (2015) Extending association rule summarization techniques to assess risk of diabetes mellitus. IEEE Trans Knowl Data Eng 27(1):130–141
14. Sirisomboonrat S, Sinapiromsaran K (2012) Breast cancer diagnosis using multi-attributed lens recursive partitioning algorithm. In: Tenth International Conference on ICT and Knowledge Engineering, IEEE
15. Sousa T et al (2004) Particle Swarm based data mining algorithms for classification tasks. Parallel Comput—Special issue: Parallel and nature-inspired computational paradigms and applications 30(5–6):767–783
16. Tripoliti EE et al (2012) Automated diagnosis of diseases based on classification: dynamic determination of the number of trees in random forests algorithm. IEEE Trans Inf Technol Biomed 16(4):615–622

Chapter 4
QSAR and Validation Analysis on MMP-13 Inhibitors

G. Nirmala, Y.B. Adimulam and P. Seetharamaiah

Abstract Matrix metalloproteinases (MMPs) play an important role in the tissue modeling and remodeling of the extracellular matrix. Abnormal activity of these enzymes was associated with metastasis, angiogenesis, cardiovascular disease, osteoarthritis, and rheumatoid arthritis. Multiple regression procedure was employed to perform QSAR (Quantitative Structure Activity Relationship) analysis on a set of 72 α-sulfone hydroxamate MMP-13 inhibitors. Outlying data was removed using Relative Error calculation and Extent of Extrapolation. The activity contributions of 57 compounds after removing outliers, was determined and the validation procedures such as cross-validation r^2, R^2ev, ext etc. were obtained. The generated model could be useful in designing more potent inhibitors of MMP-13.

Keywords Sulfone hydroxamates · QSAR · Multiple regression · MMP-13

4.1 Introduction

Matrix metalloproteinase (MMPs) play an important role in the tissue modeling and remodeling of the extracellular matrix in both physiologic and pathologic states and thus plays an important role in tumor progression [1, 2]. Matrix metalloproteinases are structurally similar, but differ in substrate specificity, in that each MMP has the ability to degrade particular subset of matrix proteins [3]. Abnormal activity of these enzymes has been related to a variety of pathologic processes, involving metastasis, angiogenesis, cardiovascular disease, osteoarthritis, and rheumatoid arthritis [4]. The development of potent subclass selective inhibitors of these enzymes has been challenging, and they rely on a small number of zinc binding motifs [5].

G. Nirmala · Y.B. Adimulam (✉)
Sir C.R. Reddy College of Engineering, Eluru, India
e-mail: yesubabuadimulam9@gmail.com

P. Seetharamaiah
Department CS and SE, Andhra University, Visakhapatnam, India

© The Author(s) 2016
P.V. Lakshmi et al. (eds.), *Computational Intelligence Techniques
in Health Care*, Springer Briefs in Forensic and Medical Bioinformatics,
DOI 10.1007/978-981-10-0308-0_4

MMP inhibitors bind to members of ADAMs family (A Disintegrin and Metalloprotease), [6] leading to joint problems. Therefore, another approach to design selective MMP inhibitors is to optimize the selective inhibition of the single MMP isozyme, reducing other protease inhibitions [7, 8]. MMP-13 rapidly degrades type II collagen and is is upregulated in osteoarthritis joints and in cancer [9].

Therefore in search of potent MMP-13 inhibitors, a novel series of various α-sulfone hydroxamates reported in three papers [7, 8, 10, 11] were considered to perform structure-activity relationship studies.

QSAR studies have been investigated on the basis of the fact that the biological activity of the compound is a function of its physicochemical properties. From literature it was observed that several attempts were made to build QSAR models of various MMP inhibitors such as, non-zinc chelating compounds [12, 13], piperazine analogs [14], carboxylic acid based compounds [12, 13], N-hydroxy-a-phenylsulfonylacetamide [15] and docking based QSAR [16]. Moreover, none of the QSAR studies reported on a-sulfone hydroxamate analogs that covered two or more different kinds of ligands. Hence, a QSAR study on ligands with observable structure diversity, if possible, will definitely lead to more universal and robust QSAR models for designing novel compounds against MMP-13.

4.2 Materials and Methods

4.2.1 Data Set

A data set of 72 α-sulfone hydroxamate compounds [7, 8, 10, 11] reported in literature was utilized to obtain a reliable and robust QSAR model. The IC50 inhibitory activities of these derivatives and their 2-dimensional structures are given in Table 4.1. The structures were drawn using ISIS Draw 2.3 (www.mdli.com) software and the descriptors were calculated using Tsar Software. Before calculation of descriptors, three dimensional structures of all molecules was generated, charges derived and the geometries were optimized.

4.2.2 Regression Analysis

QSAR model was constructed on complete data set. Validation was done using leave-one-out (LOO) technique. The relationship between dependent variable (log1/IC50) and independent variables was established by linear multiple regression analysis. Significant descriptors for QSAR equation were chosen based on the statistical data like correlation coefficient (r), standard error of estimate (s), F-value, cross-validation r2 (q2) and predictive residual sum of squares (PRESS). Cross-validation was calculated using 2 random trials with F to leave and F to enter being 2 in F stepping to include the most significant variables.

Table 4.1 Structures and biological activities of a-sulfone hydroxamate derivatives as MMP-13 inhibitors

ID	X	NR^{1}R^2	IC$_{50}$ (nM)	Log 1/IC$_{50}$
1	O	Allyl(methyl)amino	35.0	−1.544
2	O	Methyl(prop-2-ynyl)amino	24.5	−1.389
3	N-cyclopropyl	Benzyl(methyl)amino	20.0	−1.301
4	O	3,4-Dihydroisoquinolin-2(1H)-yl	9.0	−0.954
5	O	6,7-Dimethoxy-3,4-dihydroisoquinolin-2 (1H)-yl	6.2	−0.792
6	O	3,5-Dimethylpiperidin-1-yl	4.4	−0.643
7	N-CH$_2$CH$_2$OMe	3,5-dimethylpiperidin-1-yl	50.0	−1.699
8	O	cis-2,6-Dimethylmorpholin-4-yl	18.1	−1.258
9	O	4-Acetylpiperazin-1-yl	50.0	−1.699
10	O	4-Isopropylpiperazin-1-yl	28.0	−1.447
11	O	4-(2-Methoxyethyl)piperazin-1-yl	45.0	−1.653
12	O	4-Phenethylpiperazin-1-yl	25.4	−1.405
13	O	4-(2-Hydroxyethyl)piperazin-1-yl	40.0	−1.602
14	O	4-(2-(Dimethylamino)ethyl) piperazin-1-yl	90.0	−1.954
15	O	4-(2-Fluorophenyl)piperazin-1-yl	6.7	−0.826
16	O	4-(2-Methoxyphenyl)piperazin-1-yl	18.0	−1.255
17	O	4-(4-Fluorophenyl)piperazin-1-yl	6.0	−0.778
18	O	4-(4-Acetylphenyl)piperazin-1-yl	6.7	−0.826
19	O	4-(2,4-Dimethylphenyl)piperazin-1-yl	12.2	−1.086
20	O	4-(Pyridin-2-yl)piperazin-1-yl	10.7	−1.029
21	O	4-(Pyrimidin-2-yl)piperazin-1-yl)	6.4	−0.806
22	O	4-(Pyridin-4-yl)piperazin-1-yl	30.0	−1.477
23	O	4-(Pyrazin-2-yl)piperazin-1-yl	26.8	−1.428

| 24 | O | – | 4.0 | −0.602 |
| 25 | N-Cyclopropyl | – | 27.7 | −1.442 |

(continued)

Table 4.1 (continued)

26	NCH2CH2OMe	–		70	−1.845

27	NCH2CH2OMe	–		9.0	−0.954

	X	R	IC$_{50}$ (nM)	Log 1/IC$_{50}$
28	O	H	1.7	−0.230
29	O	2-F	2.7	−0.431
30	O	2-Me	7.7	−0.886
31	O	2-Cl	9.0	−0.954
32	O	2-MeO	130	−2.114
33	O	3-MeO	8.0	−0.903
34	O	3-CF$_3$	20	−1.301
35	O	4-MeO	0.63	0.201
36	O	4-Me	1.9	−0.279
37	O	2,4-diMe	28.6	−1.456
38	N-cPr	H	3.3	−0.519
39	N-cPr	4-CF$_3$	2.4	−0.380

	R^1	R^2	IC$_{50}$ (nM)	Log 1/IC$_{50}$
40	H	H	6.0	−0.778
41	MeO	H	17.5	−1.243
42	H	4-Cl	0.7	0.155
43	Cl	H	42.5	−1.628
44	Me	H	32.7	−1.515
45	CF$_3$	H	28.8	−1.459

(continued)

Table 4.1 (continued)

	R^1	R^2	IC$_{50}$ (nM)	Log 1/IC$_{50}$
46	EtO	H	35.0	−1.544
47	OH	H	11.3	−1.053
48	4-F-C$_6$H$_4$	H	33.7	−1.528
49	2,3-(CH=CH)	naphthyl	11.4	−1.057
50	Me	4-MeO	7.0	−0.845
51	MeO	4-diMeO	11.0	−1.041
52	MeO	5-diMeO	11.4	−1.057
53	MeO	5-iPr	70	−1.845

	X	R^1	IC$_{50}$ (nM)	Log 1/IC$_{50}$
54	O	CF$_3$	0.6	0.222
55	O	CL	1.0	0.000
56	cPr-N	OCH$_3$	0.42	0.377

	N	Y	IC$_{50}$ (nM)	Log 1/IC$_{50}$
57	4		0.4	0.398
58	3		1.6	−0.204
59	3		1.6	−0.204
60	3		0.66	0.180
61	3		8.0	−0.903
62	3		8.2	−0.914

(continued)

50   G. Nirmala et al.

Table 4.1 (continued)

	N	Y	IC$_{50}$ (nM)	Log 1/IC$_{50}$
63	3		8.2	−0.914
64	2		0.9	0.046
65	3		0.2	0.699
66	3		1.7	−0.230
67	3		0.3	0.523
68	3		0.2	0.699
69	3		3.3	−0.519
70	3		1.2	−0.079
71	4		1.8	−0.255
72	4		0.9	0.046

4.2.3 Molecular Descriptors

Molecular descriptors selected in the study are: topological, shape and connectivity indices, total dipole and lipole, molecular weight, h-bond donors, h-bond acceptors, logP and rotatable bond counts, heat of formation and electrostatic properties like HOMO (Highest Occupied Molecular Orbital), LUMO (Lowest Unoccupied Molecular Orbital).

4.3 Results and Discussion

Multivariate regression analysis on 72 α-sulfone hydroxamate MMP-13 inhibitors with F stepping and cross-validation by leaving-out-one row, to test the predictive power resulted in eight influential descriptors for inhibition. They are: inertia moments, lipole components, shape flexibility and 6-membered rings. Equation given below represents the linear QSAR model from a complete set of 72 inhibitors.

$$
\begin{aligned}
\log(1/IC50) = \ & +0.78127909 * \text{Inertia Moment 1 Size} \\
& +1.0273268 * \text{Inertia Moment 1 Length} \\
& -0.19030687 * \text{Total Lipole} \\
& -0.12560831 * \text{Lipole X component} \\
& -0.22934534 * \text{Lipole Z component} \\
& -0.81264067 * \text{Shape Flexibility} \\
& -0.67134702 * \text{Randic Topological index} \\
& -0.1585972 * 6 - \text{membered aliphatic rings} \\
& -0.83904165
\end{aligned}
$$

$$ r = 0.839, \quad r2 = 0.705, \quad q2 = 0.60, \quad F = 18.798, \quad n = 72, \quad s = 0.398 $$

$$(4.1)$$

4.3.1 Outlier Detection

The criterion for removing outliers is based on Relative Error calculation and Extent of Extrapolation.

4.3.1.1 Relative Error Calculation

This method was employed to calculate the relative error (Eq. 4.2) of all compounds in the data set. From Table 4.2, it cannot be stated that the model predicted

Table 4.2 Logarithmic molar concentration values of 57 molecules of the proposed QSAR model (Eq. 4.3)

ID	Activity log $(1/IC_{50})$	Predicted activity log $(1/IC_{50})$
1	−0.23	−0.3912
2	−0.431	−0.66191
3	−0.886	−0.91566
4	−0.954	−0.79564
5	−2.114	−1.20192
6	−0.903	−0.91285
7	−1.301	−1.05092
8	0.201	−0.47768
9	−0.279	−0.47494
10	−1.456	−1.06018
11	0.222	−0.4706
12	0	−0.32529
13	−0.778	−0.81823
14	−1.243	−1.27459
15	0.155	−0.61472
16	−1.628	−1.03736
17	−1.515	−0.964
18	−1.459	−1.53202
19	−1.544	−1.22372
20	−1.053	−0.91292
21	−1.528	−1.41449
22	−1.057	−0.76503
23	−0.845	−0.90006
24	−1.041	−1.44219
25	−1.057	−1.3193
26	−1.845	−1.3901
27	−0.519	−0.27506
28	−0.38	−0.11736
29	0.377	−0.16176
30	−0.204	−0.41719
31	−0.204	−0.21031
32	0.18	0.359787
33	−0.903	−0.86898
34	−0.914	−0.91161
35	−0.914	−1.08902
36	0.046	−0.28091
37	0.699	−0.13163
38	−0.23	0.462171
39	0.523	0.505681
40	0.699	1.15629
41	−0.519	−0.36744

(continued)

Table 4.2 (continued)

ID	Activity log $(1/IC_{50})$	Predicted activity log $(1/IC_{50})$
42	−0.079	−0.56857
43	−0.255	−0.20775
44	0.046	0.409178
45	0.398	−0.17199
46	−1.544	−1.13317
47	−0.954	−1.1284
48	−1.389	−0.97358
49	−1.301	−1.22672
50	−0.954	−1.07234
51	−0.792	−0.92055
52	−0.643	−1.44345
53	−1.699	−1.86919
54	−1.258	−1.33341
55	−1.699	−1.3192
56	−1.447	−1.06291
57	−1.653	−1.49505
58	−1.4	−1.13579
59	−1.602	−1.52688
60	−1.954	−1.25951
61	−0.826	−1.04217
62	−1.255	−1.48239
63	−0.778	−0.84891
64	−0.826	−1
65	−1.086	−0.97125
66	−1.029	−1.05085
67	−0.806	−1.28563
68	−1.477	−0.91795
69	−1.428	−1.3446
70	−0.602	−1.15113
71	−1.442	−1.11378
72	−1.845	−2.06817

wrongly for the highlighted compounds, instead it can be emphasized that the model prediction led to a high relative error for compounds 8, 12, 15, 29–30, 36–38, 42, 44–45 and 52 (Table 4.3) and hence these compounds should be excluded from the study as they influence the outcome in a significant manner.

$$\text{Relative Error} = \text{Residual Value}/\text{Actual Value} \qquad (4.2)$$

Table 4.3 Relative error calculation on complete data set

S. No.	Compound No.	Actual value	Predicted value	Residual value	Relative error
1	2_2a.mol	−0.230	−0.391	0.161	−0.701
2	2_2b.mol	−0.431	−0.662	0.231	−0.536
3	2_2c.mol	−0.886	−0.916	0.030	−0.033
4	2_2d.mol	−0.954	−0.796	−0.158	0.166
5	2_2e.mol	−2.114	−1.202	−0.912	0.431
6	2_2 f.mol	−0.903	−0.913	0.010	−0.011
7	2_2 g.mol	−1.301	−1.051	−0.250	0.192
8	2_2 h.mol	0.201	−0.478	0.679	**3.376**
9	2_2i.mol	−0.279	−0.475	0.196	−0.702
10	2_2j.mol	−1.456	−1.060	−0.396	0.272
11	2_2 l.mol	0.222	−0.471	0.693	**3.120**
12	2_2 m.mol	0.000	−0.325	0.325	–
13	2_3a.mol	−0.778	−0.818	0.040	−0.052
14	2_3b.mol	−1.243	−1.275	0.032	−0.025
15	2_3c.mol	0.155	−0.615	0.770	**4.966**
16	2_3d.mol	−1.628	−1.037	−0.591	0.363
17	2_3e.mol	−1.515	−0.964	−0.551	0.364
18	2_3 f.mol	−1.459	−1.532	0.073	−0.050
19	2_3 g.mol	−1.544	−1.224	−0.320	0.207
20	2_3 h.mol	−1.053	−0.913	−0.140	0.133
21	2_3i.mol	−1.528	−1.414	−0.114	0.074
22	2_3j.mol	−1.057	−0.765	−0.292	0.276
23	2_3 k.mol	−0.845	−0.900	0.055	−0.065
24	2_3 l.mol	−1.041	−1.442	0.401	−0.385
25	2_3 m.mol	−1.057	−1.319	0.262	−0.248
26	2_3n.mol	−1.845	−1.390	−0.455	0.247
27	2_8a.mol	−0.519	−0.275	−0.244	0.470
28	2_8b.mol	−0.380	−0.117	−0.263	0.691
29	2_8c.mol	0.377	−0.162	0.539	**1.429**
30	3_12a.mol	−0.204	−0.417	0.213	**−1.045**
31	3_12b.mol	−0.204	−0.210	0.006	−0.031
32	3_12c.mol	0.180	0.360	−0.180	−0.999
33	3_12d.mol	−0.903	−0.869	−0.034	0.038
34	3_12e.mol	−0.914	−0.912	−0.002	0.003
35	3_12 f.mol	−0.914	−1.089	0.175	−0.191
36	3_13.mol	0.046	−0.281	0.327	**7.107**
37	3_14.mol	0.699	−0.132	0.831	**1.188**
38	3_15a.mol	−0.230	0.462	−0.692	**3.009**
39	3_15b.mol	0.523	0.506	0.017	0.033

(continued)

Table 4.3 (continued)

S. No.	Compound No.	Actual value	Predicted value	Residual value	Relative error
40	3_15c.mol	0.699	1.156	−0.457	−0.654
41	3_15d.mol	−0.519	−0.367	−0.152	0.292
42	3_15e.mol	−0.079	−0.569	0.490	**−6.197**
43	3_16.mol	−0.255	−0.208	−0.047	0.185
44	3_17.mol	0.046	0.409	−0.363	**−7.895**
45	3_3a.mol	0.398	−0.172	0.570	**1.432**
46	4a_1.mol	−1.544	−1.133	−0.411	0.266
47	4aa_1.mol	−0.954	−1.128	0.174	−0.183
48	4b_1.mol	−1.389	−0.974	−0.415	0.299
49	4c_1.mol	−1.301	−1.227	−0.074	0.057
50	4d_1.mol	−0.954	−1.072	0.118	−0.124
51	4e_1.mol	−0.792	−0.921	0.129	−0.162
52	4f_1.mol	−0.643	−1.443	0.800	**−1.245**
53	4g_1.mol	−1.699	−1.869	0.170	−0.100
54	4h_1.mol	−1.258	−1.333	0.075	−0.060
55	4i_1.mol	−1.699	−1.319	−0.380	0.224
56	4j_1.mol	−1.447	−1.063	−0.384	0.265
57	4k_1.mol	−1.653	−1.495	−0.158	0.096
58	4l_1.mol	−1.400	−1.136	−0.264	0.189
59	4m_1.mol	−1.602	−1.527	−0.075	0.047
60	4n_1.mol	−1.954	−1.260	−0.694	0.355
61	4o_1.mol	−0.826	−1.042	0.216	−0.262
62	4p_1.mol	−1.255	−1.482	0.227	−0.181
63	4q_1.mol	−0.778	−0.849	0.071	−0.091
64	4r_1.mol	−0.826	−1.000	0.174	−0.211
65	4s_1.mol	−1.086	−0.971	−0.115	0.106
66	4t_1.mol	−1.029	−1.051	0.022	−0.021
67	4u_1.mol	−0.806	−1.286	0.480	−0.595
68	4v_1.mol	−1.477	−0.918	−0.559	0.379
69	4w_1.mol	−1.428	−1.345	−0.083	0.058
70	4x_1.mol	−0.602	−1.151	0.549	−0.912
71	4y_1.mol	−1.442	−1.114	−0.328	0.228
72	4z_1.mol	−1.845	−2.068	0.223	−0.121

Remaining compounds 8, 12, 15, 29–30, 36–38, 42, 44–45 and 52 are disregarded from analysis

4.3.1.2   Extent of Extrapolation

Outliers should be removed in order to obtain the best statistical result [17–19]. Data set from Eq. (4.1) was selected to carry out extent of extrapolation graph. Extent of extrapolation graph plotted using MedCalc software (Fig. 4.1).

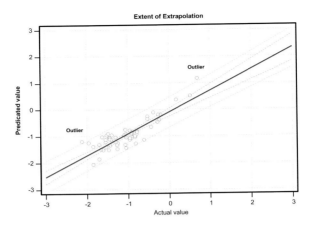

Fig. 4.1 Extent of extrapolation graph plotted using MedCalc software displaying compounds (5) and (40) as outliers (*Middle Dark line* regression line). *Dotted lines* 95 % confidence level. *Solid lines* 95 % prediction level. From the above graph, it is evidenced that all values lie within 95 % prediction levels, whereas compounds (5) and (40) fall outside the region. After extrapolation of regression line further, it was observed that the predicted activities for some of the target compounds would fall within 95 % prediction level

After removing 15 compounds as outliers, the new QSAR model was re-constructed and the equation data is given below (Eq. 4.3). From the equation and statistics [20], it is evidenced that the correlation coefficient and other values enhanced and new descriptors appeared.

$$
\begin{aligned}
\log(1/IC50) = {}& -0.045102272 * \text{Lipole Z Component} \\
& - 0.27978137 * \text{Shape Flexibility index} \\
& - 1.4712163 * \text{Balaban Topological index} \\
& - 0.33241767 * \text{Number of Cl Atoms} \\
& - 0.72112012 * 6 - \text{membered aliphatic rings} \\
& + 0.26994368 * \text{ADME H} - \text{bond Donors} \\
& - 0.20847064 * \text{ADME logP} \\
& + 4.1619339
\end{aligned}
$$

$$
r = 0.853, \quad r2 = 0.728, \quad q2 = 0.798, \quad F = 18.76, \quad n = 57, \quad s = 0.314
$$

$$(4.3)$$

A graph is plotted between observed values and predicted values, given in Fig. 4.2.

The generated QSAR model indicates that a high value of ADME H-bond Donors on MMP-13 α-sulfone hydroxamate inhibitors contributes positively to the activity, which means that a marginal increase in H-bond donating groups on compounds would enhance MMP-13 inhibition. On the other hand, a lipole Z

Fig. 4.2 Observed and predicted values of molecules in complete set after removing outliers

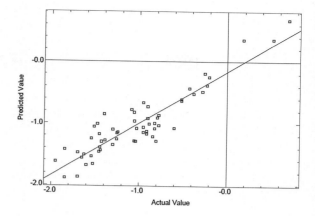

component, 3-dimensional shape flexibility of side chains, balaban topological index, Number of Cl Atoms, 6-membered aliphatic rings and logP contributes negatively to activity. Most of the studied inhibitors contain linear aliphatic groups and have the tendency to rotate the compound within the active site region. Therefore, reducing these properties on α-sulfone hydroxamates would enhance MMP-13 inhibition.

4.3.1.3   Y-Randomization

This test ensures the robustness of a QSAR model [18] and to assess the multiple linear regression models obtained by descriptor selection [17, 19]. In y-randomization test, the dependent variable or y-data is randomly shuffled and a new QSAR model is developed keeping X-data intact. The new models are expected to have low R2 and Q2 values, which determine the statistical significance of the original model (Table 4.4).

Table 4.4 R^2 and Q^2 values after several y-randomization tests

Iteration	R^2	Q^2	Iteration	R^2	Q^2
1	0.28	0.12	11	0.15	0.41
2	0.25	0.22	12	0.24	0.22
3	0.15	0.33	13	0.26	0.10
4	0.38	0.25	14	0.12	0.38
5	0.11	0.12	15	0.35	0.24
6	0.14	0.11	16	0.28	0.41
7	0.29	0.33	17	0.18	0.34
8	0.21	0.38	18	0.44	0.14
9	0.28	0.31	19	0.17	0.11
10	0.10	0.45	20	0.09	0.15

4.4 Conclusion

The QSAR model generated on a set of 57 compounds after removing outliers resulted in a promising model involving seven descriptors [Lipole Z Component, Shape Flexibility index, Balaban Topological index, Number of Cl Atoms, 6-membered aliphatic rings, ADME H-bond Donors and logP] was found to be important in enhancing MMP-13 inhibition. The model offered a useful alternative to the time consuming experiments for MMP-13 inhibition. Considering the advantages of QSAR, the model indicate that accurate predictions can be achieved with computational analysis in a reliable manner and experimental evaluation of compounds with suggested descriptors would help in designing more potent MMP-13 inhibitors.

Acknowledgments We would like to thank Dr. P. Ajay Babu, Sr. Scientist, TRIMS Labs, Visakhapatnam for his valuable suggestions and guidance on the work.

References

1. Brown PD et al (1995) Matrix metalloproteinase inhibition: A review of anti-tumour activity. Ann Oncol 6:967–974
2. Gross J et al (1962) Collagenolytic activity in amphibian tissues - a tissue culture assay. Proc Natl Acad Sci USA 48:1014–1022
3. Vihinen P et al (2005) Matrix metalloproteinases as therapeutic targets in cancer. Curr Cancer Drug Targets 5:203–220
4. Engel CK (2005) Structural basis for the highly selective inhibition of MMP-13. Chem Biol 12:181–189
5. Matter H et al (2004) Recent advances in the design of matrix metalloproteinase inhibitors. Curr Opin Drug Disc Devel 7:513–535
6. Wasserman ZR (2005) Making a new turn in matrix metalloprotease inhibition. Chem Biol 12:143–144
7. Thomas E. Barta et al (2011) MMP-13 selective a-sulfone hydroxamates: A survey of P10 heterocyclic amide isosteres. Bioorg Med Chem Lett 21:2820–2822
8. Fobian YM et al (2011) MMP-13 selective alpha-sulfone hydroxamates: identification of selective P1' amides. Bioorg Med Chem Lett 21:2823–2825
9. Johnson AR et al (2007) Discovery and characterization of a novel inhibitor of matrix metalloproteinase-13 that reduced cartilage damage in vivo without joint fibroplasis side effects. J Biol Chem 282:27781–91
10. Kolodziej Stephen A et al (2010) MMP-13 selective isonipecotamide α-sulfone hydroxamates. Bioorg Med Chem Lett 20:3561–3564
11. Kolodziej Stephen A et al (2010) Orally bioavailable dual MMP-1/MMP-14 sparing, MMP-13 selective α-sulfone hydroxamates. Bioorg Med Chem Lett 20:3557–3560
12. Sharma BK et al (2013) Chemometric descriptor based QSAR rationales for the MMP-13 inhibition activity of non-zinc-chelating compounds. Med Chem 3:168–178
13. Sharma BK et al (2013) QSAR rationale of matrix metalloproteinase inhibition activity in a class of carboxylic acid based compounds. Br J Pharm Res 3:697–721
14. Kulkarni GB et al (2009) 3D-QSAR studies of matrix metalloproteinase-13 inhibitors. Rasayan J Chem 2:407–414

15. Fernandez Michael et al (2007) QSAR modeling of matrix metalloproteinase inhibition by N-hydroxy-a-phenylsulfonylacetamide derivatives. Bioorg Med Chem 15:6298–6310
16. Xi et al (2014) In silico study combining docking and QSAR methods on a series of matrix metalloproteinase 13 inhibitors. Arch Pharm. doi:10.1002/ardp.201400200
17. Afantitis A et al (2006) A novel QSAR model for predicting induction of apoptosis by 4-aryl-4H-chromenes. Bioorg Med Chem 14:6686–6694
18. Kim D et al (2006) The quantitative structure-mutagenicity relationship of polycylic aromatic hydrocarbon metabolites. Int J Mol Sci 7:556–570
19. Afantitis A et al (2006) Investigation of substituent effect of 1-(3,3-diphenylpropyl)-piperidinyl phenylacetamides on CCR5 binding affinity using QSAR and virtual screening techniques. J Comput Aided Mol Des 20:83–95
20. Golbraikh A et al (2002) Beware of q2! J Mol Graph Model 20:269–276
21. Kubinyi H (1994) Variable selection in QSAR studies. II. A highly efficient combination of systematic search and evolution. Quant Struct Act Relat 13:393–401
22. Kubinyi H (1994) Variable selection in QSAR studies. I. An evolutionary algorithm. Quant Struct Act Relat 13:285–294

Chapter 5
Detection of Lesion in Mammogram Images Using Differential Evolution Based Automatic Fuzzy Clustering

A. Srikrishna, B. Eswara Reddy and V. Sesha Srinivas

Abstract Breast cancer is one of the major causes of cancer deaths in women. CAD (Computer Aided Detection) is becoming an integral part and helping radiologists to detect cancer as early as possible. Different CAD solutions have been devised, but we require a system that provides accurate detection of relevant lesions. This paper uses a DE based Automatic Fuzzy Clustering (DEAFC) algorithm for detection of lesions from mammogram images. The algorithm is tested on DDSM, MIAS and identifies the lesions. The performance is compared with the ground truth values which are manual markings of radiologist.

Keywords Clustering · Fuzzy clustering · Differential evolution · Mammograms

5.1 Introduction

Segmentation of medical images is a preprocessing step in radiotherapy planning. Computed Tomography (CT) and Magnetic Resonance Imaging (MRI), the two widely used radiographic techniques, produce sequences of images representing 2D sections of a 3D anatomical structure of interest [1]. Such medical imaging techniques are useful in diagnosis, clinical studies and treatment planning. Structures of

A. Srikrishna (✉) · V. Sesha Srinivas
Department of Information Technology, Rayapati Venkata Ragarao and Jagarlamudi
Chandramouli College of Engineering, Guntur, AP, India
e-mail: atlurisrikrishna@yahoo.com

V. Sesha Srinivas
e-mail: vangipuramseshu@gmail.com

B. Eswara Reddy
Department of Computer Science and Engineering,
Jawaharlal Nehru Technological University, Ananthapur, AP, India
e-mail: eswarcsejntu@gmail.com

© The Author(s) 2016
P.V. Lakshmi et al. (eds.), *Computational Intelligence Techniques
in Health Care*, Springer Briefs in Forensic and Medical Bioinformatics,
DOI 10.1007/978-981-10-0308-0_5

61

interest in medical images can be found by segmentation. Clustering is an unsupervised classification technique that aims at grouping of homogeneous elements into meaningful k clusters. The value of k is a priory in most of the clustering algorithms for example k-means, fuzzy k-means, etc., [2]. The objective of the fuzzy k-means algorithm is to maximize the global compactness of the clusters [3]. In order to overcome the limitation of the traditional clustering algorithms many heuristic optimization algorithms such as Genetic Algorithm (GA), Particle SO, DE etc. are proposed [4]. Differential Evolution (DE) is one of the most powerful stochastic real parameter optimization techniques in current use [5]. DE follows similar computational steps as in a standard Evolutionary algorithm [6–8]. Compared to other Evolutionary algorithms DE is very simple to code. The recent studies on DE provide a better performance in terms of accuracy, robustness and convergence speed with its simplicity [9–11]. The number of control parameters in DE is very few compared to other algorithms [12]. As in most clustering algorithms in DE also the number of clusters is a priory, a solution is proposed, Automatic Clustering using Differential Evolution (ACDE) [13]. The ACDE introduced a new chromosome structure to find the optimal clusters automatically. The features of DE attracted many researchers to provide more competitive solutions. A Kernel-induced fuzzy clustering using Differential Evolution is proposed [14]. Evolution of clusters using point symmetry method is proposed and they have used a point symmetry based cost function as objective function [15]. Proved the efficiency of ACDE in finding tissues in medical images [16], the segments generated are more nearer to the markings of the radiologists. If accuracy is the only prime factor then, in most of the real applications fuzzy performs well compared to k-means. Hence, the DE based Automatic Fuzzy clustering is proposed by Das and Konar [17]. With the observed improvement of the algorithm we confine to apply on medical images especially to detect shapes of tissues in the images.

This paper is organized as follows. The existing system for segmentation of images Fuzzy-k-means algorithm is discussed in Sect. 5.2. The proposed method DE based Automatic Fuzzy Clustering is discussed in Sect. 5.3. Performance of the algorithm is discussed in Sect. 5.4 and finally conclusion is presented in Sect. 5.5.

5.2 Fuzzy-k-Means Algorithm

In this method a data point can belong to several groups with a membership degree between 0 and 1 which is shown by a matrix U_{kXm} where k is number of clusters and m is number of data points. U_{kXm} can be computed using Eq. 5.1.

$$\sum_{r=1}^{k} U_{ri} = 1 \quad \forall \quad i = 1, 2, \ldots, m \tag{5.1}$$

where U_{ri} is the membership degree of ith pattern in rth cluster. The cost function used in this method is shown in Eq. 5.2, where p is the fuzzification factor:

$$J(U, c_1, \ldots, c_k) = \sum_{r=1}^{k} \sum_{i=1}^{m} u_{ri}^p d(x_i, c_r) \qquad (5.2)$$

The membership of each data item and cluster centres are recomputed with the Eqs. 5.3 and 5.4 respectively to minimise the cost function.

$$u_{ri} = \frac{1}{\sum_{i=1}^{k} \left(\frac{d(x_i, c_r)}{d(x_i, c_l)}\right)^{2/(p-1)}} \qquad (5.3)$$

$$c_r = \frac{\sum_{i=1}^{m} u_{ri}^p x_i}{\sum_{j=1}^{m} u_{rj}^p} \qquad (5.4)$$

where x_i is the ith element in the dataset. The fuzzy k-means algorithm and k-means algorithms are very similar in processing.

5.2.1 Steps of the Algorithm

1. Choose the number of clusters.
2. Assign membership to each data point being in the clusters randomly.
3. Repeat the step 4 and step 5 until the algorithm converges. The converging condition is that the change in the cost function of the two consequent iterations is not more than the given sensitivity threshold.

   i. Using the formula (5.4), the centroids are computed for each cluster.
   ii. Reassign the memberships to each of the data point being in the cluster using the formula (5.3).

The algorithm minimizes intra-cluster variance, but has the same problems as k-means; the minimum is a local minimum, and the results depend on the initial choice of seeds.

5.3 Differential Evolution Based Automatic Fuzzy Clustering (DEAFC)

The classical DE [5], is a population-based global optimization algorithm that uses a floating-point (real-coded) representation. The details of Differential Evolution are as follows.

A population is a set of solutions represented as chromosomes (vectors). The population is represented at different timestamps called generations. The chromosome i of a population at timestamp t is represented as

$$\overrightarrow{Z_i}(t) = \left[Z_{i,1}(t), Z_{i,2}(t), \ldots, Z_{i,d}(t)\right]$$

For each chromosome Z_k of the population in the current generation, the algorithm (DE) randomly finds three individual chromosomes (Z_i, Z_j, Z_m,) from the current generation called parents. After finding the parents, it calculates the difference of Z_i and Z_j and scales it by a scalar factor F [0, 1] creates new chromosome called offspring by adding the result to Z_m. Using Eq. 5.5, DE decides whether the Z_k or $Z_m + F(Z_i - Z_j)$ will be in the next generation. Here Cr is called cross over rate is a scalar parameter value ranges between 0 and 1.

$$U_{k,n}(t+1) = \begin{cases} Z_{m,n} + F\left(Z_{i,n}(t) - Z_{j,n}(t)\right) & , if\ rand_n(0,1) < Cr \\ Z_{k,n}(t), otherwise \end{cases} \quad (5.5)$$

If the new offspring yields a better value of the objective function, then it replaces the Z_k in the next generation otherwise, the Z_k is reserved in the population using Eq. 5.6.

$$\overrightarrow{Z_i}(t+1) = \begin{cases} \overrightarrow{U_i}(t+1), if\ f\left(\overrightarrow{U_i}(t+1)\right) > f\left(\overrightarrow{Z_i}(t)\right) \\ \overrightarrow{Z_i}(t), if\ \overrightarrow{U_i}(t+1) \leq f\left(\overrightarrow{Z_i}(t)\right) \end{cases} \quad (5.6)$$

where $f(\cdot)$ is the objective function to be maximized. In the original DE, the difference vector $(Z_i(t) - Z_j(t))$ is scaled by a F called constant factor. The control parameter F is a number between 0.4 and 1. Equation 5.7 is used to change the scale factor F in a random manner within the range (0.5, 1).

$$F = 0.5 * (1 + rand(0,1)) \quad (5.7)$$

where rand(0, 1) is a uniformly distributed random number within the range [0, 1]. The scale factor is computed for 100 runs and found the mean as 0.75.

5.3.1 DE Based Automatic Fuzzy Clustering (DEAFC) Algorithm

The pseudo code for the complete Fuzzy ACDE algorithm is as follows.

Step 1   Each chromosome of the first population is initialized with K number of cluster centres randomly selected and K activation thresholds in the range [0, 1] (randomly chosen).

Step 2 For each chromosome find the active cluster centres based on the activation thresholds.

Step 3 For $t = 1$ to t_{max} do

   (a) The distance $d(Xp, m_{i,j})$, from each date vector Xp to all the active cluster centres $m_{i,j}$ is calculated, and find belongingness in terms of membership of the ith chromosome V_i.

   (b) The data vector Xp is assigned to the particular cluster centre $m_{i,j}$ *based on membership in the Fuzzy.*

   (c) If any cluster centre m_{ij} is having lesser than two data vectors as members then the centroid can be updated as follows:

      i. Let L be the number of elements, where m elements equally distributed into k centres. i.e., L = m/k.

      ii. The update the centroid as the average of randomly selected L data elements.

   (d) Change the population members according to the DE algorithm outlined in the above equations. Use the cost function in Fuzzy-k as fitness of the chromosomes to guide the evolution of the population.

Step 4 The final solution is the cluster centres in the best chromosome having the highest value of the fitness function at time $t = t_{max}$.

5.4 Experimental Results

The performance of DE based Automatic Fuzzy clustering has been studied on datasets DDSM and MIAS. Table 5.1 shows the improved performance of DEAFC over ACDE in determining shape of the tissue. The numeric figures associated with each segmented image is the specification of the resulted number of segments for the given input number of segments. From Table 5.1 the DEAFC produces less number of segments compared to ACDE and also it is observed that the abnormalities in the breast images are more clearly segmented. In Image 1, the ACDE results 11 out of 20 groups whereas the DEAFC finds 8 from the same input group. In Image 2, the ACDE results 18 out of 30 groups whereas the DEAFC finds 11 from the same input group. In Image 3, the ACDE results 17 out of 30 groups whereas the DEAFC finds 10 from the same input group. In Image 4, the ACDE results 19 out of 30 groups whereas the DEAFC finds 12 from the same input group. In Image 5, the ACDE results 8 out of 15 groups whereas the DEAFC finds 5 from the same input group. In Image 6, the ACDE results 29 out of 45 groups

Table 5.1 Performance of DEAFC

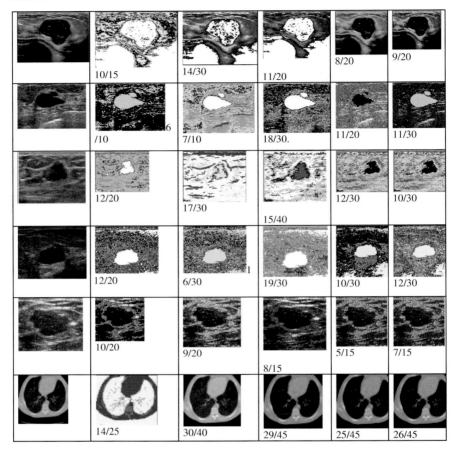

whereas the DEAFC finds 25 from the same input groups 45. We have extended the study by comparing the extracted shapes of tissues with shapes specified by a radiologist and with the chain optimization [18]. Table 5.2 shows the sample of original images of breast cancer, tissue segmentation (green) as specified by the radiologist and the corresponding tissue segmentation (red) results of Chain optimization. It is clearly observed that the segmented portions of the proposed algorithm are nearly equal to the radiologist markings.

Table 5.2 Tissue segmentation (*green*) as specified by the radiologist

S.No.	Original	Radiologist Markings	Fuzzy ACDE
1			
2			
3			

5.5 Conclusion

The performance of DEAFC has been evaluated using DDSM and MIAS images. It has been observed that DEACC gives more accurate lesions as compared to ACDE. DEAFC segmentation is nearer to the actual radiologist markings and it can be suggested for medical diagnosis. Future research may focus on employing other improved cluster validity indices to form the fitness function and a multi objective DE. Besides the pixel intensity alone, it could be interesting to take into account other features related to texture, shape and color for the segmentation task by DEAFC.

References

1. Maulik U (2009) Medical image segmentation using genetic algorithms. Inf Technol Biomed IEEE Trans 13(2):166–173
2. Jain AK, Dubes RC (1988) Algorithms for clustering data. Prentice Hall, NJ, p 320. ISBN: 013022278X

3. Bezdek JC (1981) Pattern recognition with fuzzy objective function algorithms. Plenum, New York
4. Das S et al (2009) Metaheuristic clustering. Springer, Berlin. ISBN: 978-3-540-92172-1, ISSN: 1860949X
5. Price KV et al (2005) Differential evolution—a practical approach to global optimization. Springer Natural Computing Series
6. Price KV (1997) Differential evolution vs. the functions of the 2nd ICEO. In: Proceeding IEEE international conference on evolutionary computation, pp 153–157
7. Price KV, Storn R (1997) Differential evolution: a simple evolution strategy for fast optimization. Dr. Dobb's J 22(4):18–24
8. Feoktistov V (2006) Differential evolution: in search of solutions. Springer, Berlin
9. Price KV (1999) An introduction to differential evolution. In: Corne D, Dorigo M, Glover V (eds) New ideas in optimization. McGraw-Hill, London, pp 79–108
10. Storn R, Price KV (1995) Differential evolution: a simple and efficient adaptive scheme for global optimization over continuous spaces ICSI, USA. Tech. Rep. TR-95-012
11. Storn R (2008) Differential evolution research: trends and open questions. In: Chakraborty UK (ed) Advances in differential evolution. Springer, Berlin, pp 1–32
12. Das S, Suganthan PN (2011) Differential evolution: a survey of the state-of-the-art. IEEE Trans Evol Comput 15(1):4–32
13. Das S, Abraham A (2008) Automatic clustering using an improved differential evolution algorithm. IEEE Trans on Syst Man Cybern Part A Syst Hum 38(1):218–237
14. Das S, Sil S (2010) Kernel-induced fuzzy clustering of image pixels with an improved differential evolution algorithm. Inf Sci 180:1237–1256
15. Bandyopadhyay S, Saha S (2008) A point symmetry-based clustering technique for automatic evolution of clusters. IEEE Trans Knowl Data Eng 20(11):1441–1457
16. Pavan KK et al (2012) An automatic tissue segmentation in medical images using differential evolution. J Appl Sci 12(6):587–592
17. Das S, Konar A (2009) Automatic image pixel clustering with an improved differential evolution. Appl Soft Comput 226–236
18. Rahnamayan S, Mohamad ZS (2010) Tissue segmentation in medical images based on image processing chain optimization. IEEE NPSS (Toronto), UOIT, Oshawa

Chapter 6
Sequence Alignment by Advanced Differential Evolutionary Algorithm

Lakshmi Naga Jayaprada Gavarraju, Jeevana Jyothi Pujari and K. Karteeka Pavan

Abstract In Computational biology, Biological sequence alignment plays an essential role in gene structure/function prediction. Sequence alignment is a problem of optimization that finds optimal arrangement of sequences by maximizing the similarities of residues. Metaheuristics are the modern search and optimization techniques providing competitive solutions for many real world problems. However the techniques suffer from the efficient evolution operators and control parameters. Differential Evolution is a stochastic real parameter optimization technique with few control parameters. This paper proposes an Advanced Differential Evolution (ADE) algorithm with a new mutation operator "ADE/best-worst-rand/1" considering least and best fittest candidate solutions. The performance of the algorithm is evaluated using various data sets and compared with other evolutionary algorithms, Genetic Algorithm (GA) and Differential Evolution. Experimental results have shown the efficiency and prominence of new proposed algorithm in producing best solutions for the sequence alignment with improved fitness. It is observed that performance improvement of ADE over DE is nearly 1.22 and 16.32 % over GA.

Keywords Sequence alignment · Fitness · DE · GA · Mutation

L.N.J. Gavarraju (✉)
Department of Computer Science and Engineering, Narasaraopeta Engineering College, Narasaraopet 522 601, Andhra Pradesh, India
e-mail: njayaprada@yahoo.co.in

J.J. Pujari
Department of Computer Science and Engineering, Vasireddy Venkatadri Institute of Technology, Nambur 522 509, Andhra Pradesh, India
e-mail: jyothi915@gmail.com

K. Karteeka Pavan
Department of Information Technology, R.V.R & J.C. College of Engineering, Guntur 522 019, Andhra Pradesh, India
e-mail: karteeka@yahoo.com

© The Author(s) 2016
P.V. Lakshmi et al. (eds.), *Computational Intelligence Techniques in Health Care*, Springer Briefs in Forensic and Medical Bioinformatics, DOI 10.1007/978-981-10-0308-0_6

69

6.1 Introduction

The face of biology [1] has been altered by the emergence of bioinformatics. Large scale genome sequencing research produced a very huge amount of data. The need to analyze the huge data is becoming a challenge. It demands for sophisticated methods to analyze biological sequences which are explosively expanding computational molecular biology research area and of which sequence analysis is one of the key issue [2]. Sequence analysis plays a crucial role in the study of molecular evaluation, protein structure/function prediction, RNA folding, homology of sequences, gene regulation and primer design for (PCR) polymer ace chain reaction. The sequence alignment is an arrangement of sequences of DNA/RNA or proteins to identify the regions of similarity. The number of matches of residues among the sequences is called similarity. Sequence alignments can be global or local. Global alignment maximizes the number of matched residues for the entire sequence. Local alignment maximizes similar sub regions. Sequence alignment applied only on two sequences is pair wise. An extension to pair wise is multiple sequence alignment is the alignment of three or more sequences [3–5]. The tradeoff between accuracy and efficiency is the challenging issue in aligning sequences [6]. Several methods have been proposed for sequence alignment problem. These can be categorized into classical Dynamic programming, progressive and evolutionary algorithms. Needleman–Wunsch algorithm [7] is a technique for aligning the sequences globally by maximizing the number of matches and minimizing the number of gaps for optimal alignment of sequences. DP works based on scoring scheme and generates a matrix for all possible alignments between the sequences [8, 9]. It starts at the end of the sequences and attempts to match the possible residues and giving a separate score for match, mismatch and gaps. Highest score shows the optimal alignment. However, these algorithms tend to be slow. An extensive research work is done using evolutionary optimization algorithms for finding the near optimal solution instead of best alignment [10, 11]. There have been various optimization techniques for the problem of sequence alignment. Among these genetic algorithms, simulated annealing, differential evolution and particle-swarm optimization methods are widely used because of their simplicity, versatility, and robustness [12].

Genetic algorithm (GA) are the class of evolutionary algorithms based on the concept of natural biological evolution. Genetic algorithms are suited for nature driven problems. GAs creates initial population of random solutions represented as chromosomes. On the possible solutions, GAs applies crossover and mutation operators in the search space to generate off springs. Good solutions are refined based on the principle of survival-of-the-fittest through selection process using fitness as a measure [13–16]. However, GAs suffers from premature convergence problem. Ant colony and Particle swarm optimization are also random search methods for producing solutions to sequence alignment problems [17].

Differential evolution (DE) is one of the most powerful stochastic, heuristic and population based evolutionary algorithm for generating the better solutions to global optimization problems [18, 19]. Simplicity, efficiency and ease to use are the benefits of differential evolution algorithm. Differential evolution (DE) algorithm runs through the steps as same as in standard evolutionary algorithm. In DE, random initial population is created and then improved by the operators like selection, mutation and cross over. The stopping condition for DE is the maximum number of generations G_{max} to be computed and the restricted parameters are crossover factor C_R and scaling or mutation factor F. In DE all solutions have the same probability of being selected as parents. Differential evolution algorithm utilizes difference of the parameter vectors to analyze the objective function; however experiences the difficulties in operators developed [20]. This paper proposed a new mutation operator to enhance the performance of DE proved by applying to pair wise sequence alignment problem. It is observed that performance improvement of ADE over DE is nearly 1.22 and 16.32 % over GA. This paper consists of 6 sections. Section 6.1 is introduction. Section 6.2 is containing preliminaries and definitions. Section 6.3 is explaining the methodology and the proposed mutation variant. Section 6.4 is related to experimental results. Section 6.5 is about conclusion.

6.2 Preliminaries

Pair wise alignment is aligning only two sequences, in which gaps substitutes in the same column for the like characters in a way to maximize the number of matches with respect to characters. Mismatch is considered as non-alike characters in the same column [cosine functions]. {A, C, G, T} are the residues for nucleotides, {A, C, G, U} for RNA and for the proteins, amino-acids 20-letter symbols are {A, R, N, D, C, E, F, Q, G, H, I, J, L, K, M, P, S, T, W, Y, V}. The computational techniques employs scoring function to assess the similarity of sequences by means of maximizing the number of matches and gaps might be minimized.

Definitions

The definitions and notations used throughout the paper as follows.

(i) Sequence: It is a string by concatenation of zero or more elements from an alphabet represented as Σ, sequence is considered as S.
(ii) Space or gap: It is an extra element which is not in Σ, and it is a symbol represented as—$\notin \Sigma$, but not a character.
(iii) Sequence length: The number of elements in a sequence and cardinality of S, as |S|.
(iv) Alignment: The given two sequences are $P = P_1 P_2 \ldots P_m$ and $Q = Q_1 Q_2 \ldots Q_n$, an alignment of both the sequences is $P^T = P_1^T P_2^T \ldots P_S^T$ and $Q^T = Q_1^T Q_2^T \ldots Q_S^T$ can be constructed by inserting zero or more gaps (–) into P and Q so that at least one P_i^T maps to one Q_i^T.
(v) Optimal alignment: The maximization of number of matches.
(vi) Scoring schema:

Match: $P_i^T = Q_i^T \neq -$
Mismatch: $(P_i^T \neq Q_i^T)$ and $(P_i^T, Q_i^T \neq -)$
Insertion or deletion: either P_i^T or Q_i^T is $-$.

Sum-of-pairs (SPS) function as a measure of evaluating the alignment of sequences.

6.3 Methodology

Differential evolution (DE) is simple but effective evolutionary algorithm for many optimization problems in real-world applications.

6.3.1 Differential Evolution

The Pseudo-code for Differential Evolution algorithm with Binomial Crossover is as shown below.

Step 1: Read values of the control parameters of Differential Evolution: scale factor F, crossover rate Cr, and the population size NP from user.

Step 2: Set the generation number $G = 0$ and randomly initialize a population of NP individuals $P_G = \{X_{1,G}, \ldots, X_{NP,G}\}$ with $X_{i,G} = [x_{1,i,G}, x_{2,i,G}, x_{3,i,G}, \ldots, x_{D,i,G}]$ and each individual uniformly distributed in the range $[X_{min}, X_{max}]$, where

$$X_{min} = \{x_{1,min}, x_{2,min}, \ldots, x_{D,min}\} \text{ and } X_{max} = \{x_{1,max}, x_{2,max}, \ldots, x_{D,max}\}$$

with $i = [1, 2, \ldots, NP]$ and D is the dimension of the problem.

Step 3: WHILE the stopping criterion is not satisfied
DO
FOR $i = 1$ to NP //do for each individual sequentially

Step 3.1 **Mutation Step**
Generate a donor vector $V_{i,G} = \{v_{1,i,G}, \ldots, v_{D,i,G}\}$ corresponding to the ith target
vector $X_{i,G}$ via the differential mutation scheme of DE as

$$V_{i,G} = X_{r1,G} + F \cdot (X_{r2,G} - X_{r3,G}).$$

Step 3.2 **Crossover Step**
Generate a trial vector $U_{i,G} = \{u_{1,i,G}, \ldots, u_{D,i,G}\}$ for the ith target vector $X_{i,G}$ through binomial crossover in the following way:

$u_{j,i,G} = v_{j,i,G}$, if ($rand_{i,j}[0, 1] \leq Cr$ or $j = j_{rand}$) where j_{rand} is random number in the range a and b and newly generated for each j. $x_{j,i,G}$, otherwise.

Step 3.3   **Selection Step**

Evaluate the trial vector $U_{i,G}$

IF $f(U_{i,G}) \leq f(X_{i,G})$, THEN $X_{i,G+1} = U_{i,G}$

ELSE $X_{i,G+1} = X_{i,G}$.

END IF

END FOR

Step 3.4   Increases the Generation Count

$G = G + 1$

END WHILE

In the mutation step several mutation variants are used. The mutation variants are as follows.

"DE/rand/1".

$$V_{i,G} = X_{r1,G} + F.\left(X_{r2,G} - X_{r3,G}\right)$$

"DE/best/1".

$$V_{i,G} = X_{best,G} + F.\left(X_{r1,G} - X_{r2,G}\right)$$

"DE/current-to-best/1".

$$V_{i,G} = X_{i,G} + F.\left(X_{best,G} - X_{i,G}\right) + F.\left(X_{r1,G} - X_{r2,G}\right)$$

"DE/best/2".

$$V_{i,G} = X_{best,G} + F.\left(X_{r1,G} - X_{r2,G}\right) + F\left(X_{r3,G} - X_{r4,G}\right)$$

"DE/rand/2".

$$V_{i,G} = X_{r1,G} + F.\left(X_{r2,G} - X_{r3,G}\right) + F.\left(X_{r4,G} - X_{r5,G}\right)$$

where $r_1, r_2, r_3, r_4, r_5 \in [1, \cdots, NP]$ are   randomly   chosen   integers,   and $r_1 \neq r_2 \neq r_3 \neq r_4 \neq r_5 \neq i$. F is the scaling factor controlling the amplification of the differential evolution. $X_{best,G}$ is the best individual vector with the best value in the population at generation G.

Among all the variants **"DE/rand/1"** was proved as best. Now this paper proposes a new mutation variant **"ADE/best-worst-rand/1"** by considering best, worst and random vectors from the population at generation G. And this is proved as better than **"DE/rand/1"**.

6.3.2   Proposed Mutation Variant "ADE/Best-Worst-Rand/1" for ADE

Advanced Differential Evolution algorithm is similar to Differential Evolution except in mutation strategy. The newly proposed mutation variant is as follows. **"ADE/best-worst-rand/1".**

$$\mathbf{V_{i,G}} = \mathbf{X_{r1,G}} + \mathbf{p}(X_{best,G} - X_{i,G}) + \mathbf{q}(X_{worst,G} - X_{i,G})$$

where r1 is random number between 1 and NP, X_{best} is best fit chromosome, X_{worst} is worst fit chromosome, p and q are two random numbers between [0,1].

6.3.3   Chromosome Representation

The chromosomes are generated by encoding the sequences. The gap positions in the sequence are being used to represent a chromosome. The gap positions of all the sequences are used to make a single chromosome.

Input Sequences
**>MMVHLTPMMKSAVTALWGKVNVNDGVDMVGGMALGRLLVVYPW
TQRFFMSFGDLSTP   >MMGLSDGMWQLVLMADIPGHGQMVLIRLFK
GHPMTLMKFDKFKHLKSMDMMK**
    Chromosome Representation for the selected example is as follows:

```
population(:,:,1) =

    6    68    12    23    49    21    29    22    28    64    16     0     0     0     0     0
    9    24    47    18    65    26    38    67    60     4    31    50    37    21    27    25


population(:,:,2) =

   14    43    10    11    62    57    47    26    42    35     7     0     0     0     0     0
   18    63    46     4    15    21    49    37    25    38    17    47     9    27    58    52


population(:,:,3) =

   61    29    39    14     3    36    50    41    13    26     4     0     0     0     0     0
   58    18    34    48    11    26    27     5    38    60    43    37    35    21    22     7


population(:,:,4) =

   43    57    44    60    40     2     4    37    25    23    28     0     0     0     0     0
   60    32    42    37    65     8    20    50    39    45    13    67     5    33    51    48


population(:,:,5) =

   20    28    54    66    53    63    65     9    14    23    37     0     0     0     0     0
   24    35    63     7    44     8    64    36    38     9    47     2    58    25     1    34
```

Gaps are inserted at the specified positions which are given in the chromosome representation. Now the sequences are as follows.

```
align1(:,:,1) =

MMVHL-TPMMK-SAV-TALW---GKVN--VNDGVDMVGGMALGRLLVV-YPWTQRFFMSFGDL-STP-
MMG-LSDG-MWQLVLMA-DI-PG----HGQ-MVLIR--LFKGHPMT-LM-KFDKFKHLK-SMDM-M-K

align1(:,:,2) =

MMVHLT-PM-|-MK-SAVTALWGKVN-VNDGVDMV-GGMALG--RLL-VVYPWTQRF-FMSF-GDLSTP
MMG-LSDG-MWQLV-L--MA-DIP-G-HGQMVLIRL--FKGHPMT--L-MK-FDKFK-HLKS-MDMMK

align1(:,:,3) =

MM--VHLTPMMK--SAVTALWGKVN-VN-DGVDMV-GG-M-ALGRLLVV-YPWTQRFFMS-FGDLSTP
MMGL-S-DGM-WQLVLM-AD--IPG--HGQMVL--I--RLFK-GHPM-TLMKFDKFK-H-LKSMDMMK

align1(:,:,4) =

M-M-VHLTPMMKSAVTALWGKV-N-VN-DGVDMVGG-MA-LG--RLLVVYPWTQRF-FM-SFGDLSTP
MMGL-SD-GMWQ-LVLMAD-IPGHGQMVLIR--LFK-G-HP-MT-LM-K--FDKFKHLK-SMDM-M-K

align1(:,:,5) =

MMVHLTPM-MKSA-VTALW-GK-VNVN-DGVDMVGG-MALGRLLVVYPWTQR--FFMSFGDL-S--TP
--MMGL---SDGMWQLVLMADIP--GHGQMVLI---R-LFKGH-PM-TLMKFDKFKH-LKSM--DMMK


fit =

 -156  -150  -144  -140  -152
```

6.3.4   Fitness Function

This fitness function determines how "best" an alignment is. Fitness calculation methods play an important role in the performance of evolutionary algorithms. The most common function that is used with substantial deviations is called the "Sum-Of-Pair" Objective Function. In this method, for each location on the aligned sequences, one of three situations will occur: match, mismatch or a gap. The fitness of an alignment is calculated as

$$\textit{Fitness value for n sequences} = \sum_{i=1}^{Length} (\textit{Fitness Score for each pair of elements in the column})$$

where match = 2, mismatch = −1 and gap = −2.

6.3.5   Termination Criteria

The algorithm is terminated after the defined numbers of generations.

6.4   Experimental Results

The performance of the proposed mutation vector of ADE is studied by taking 4 sets of synthetic test data sets. Each contains two sequences of variable length. They are as follows.

Test Data Set I
ACGATCGACGGATCG
GACCATCCGAT
Test Data Set II
ATCGACGACAAGGCCATACATCGACG
CAATCGACGGCCTAGACTTAC
Test Data Set III
ACTGCCAGACTCGCGACATCGAGAT
CTATCGACTGACCGTCGGA
Test Data Set IV
AGCTACGTAAGGCCATCGACGACTACGAATTCGGATGCCATCGATCG
CGATTCCGCATCGAGGCCATCGATTGGCACTGACGTATCG

The performance of the GA, DE and ADE algorithms are demonstrated in the form of Graphs and Tables as shown below.

In Test Data Set I, Seq 1 length is 15 and Seq 2 length is 11. Average fitness of GA is 1.9, Average fitness of DE is 11.078 and Average fitness of ADE is 11.126 (Fig. 6.1; Table 6.1).

In Test Data Set II, Seq 1 length is 26 and Seq 2 length is 21. The average performance improvement of ADE over GA is 16.32 % and improvement of ADE over DE is 1.22 % (Fig. 6.2; Table 6.2).

In Test Data Set III, Seq 1 length is 25 and Seq 2 length is 19. The average fitness of all the generations is 3.9 for GA, 17.84 for DE and 17.88 for ADE (Fig. 6.3; Table 6.3).

In Test Data Set IV the length of the Seq 1 is 47 and Seq 2 length is 40. Average fitness of GA is 1.7, average fitness of DE is 13.76 and average fitness of ADE is 14.36 (Fig. 6.4; Table 6.4).

Performance testing on ADE with various crossover rates

In general the crossover rate must be between 0 and 1. We tested the ADE algorithm with different crossover rates and we found that with crossover rate 0.2 the algorithm gave very good results (Fig. 6.5; Table 6.5).

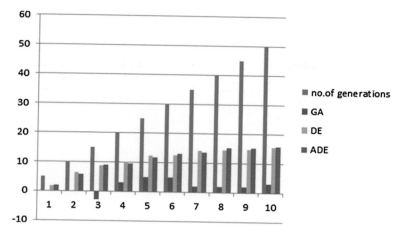

Fig. 6.1 Performance evaluation of GA, DE and ADE for Test Data Set I

Table 6.1 Fitness values of GA, DE and ADE for Test Data Set I

No. of generations	GA	DE	ADE
5	0	1.8	1.98
10	0	6.3	5.8
15	−3	8.82	9.02
20	3	10.1	9.62
25	5	12.28	11.84
30	5	12.58	13.14
35	2	14.04	13.68
40	2	14.56	15.16
45	2	14.76	15.32
50	3	15.54	15.7

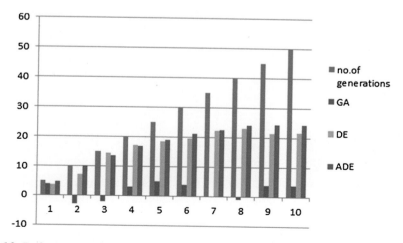

Fig. 6.2 Performance evaluation of GA, DE and ADE for Test Data Set II

Table 6.2 Fitness values of GA, DE and ADE for Test Data Set II

No. of generations	GA	DE	ADE
5	4	3.6	4.8
10	−3	7	10
15	−2	14.4	13.6
20	3	17.2	17
25	5	18.6	19
30	4	19.6	21.2
35	0	22.4	22.6
40	−1	23	24.2
45	4	21.4	24.4
50	4	21.8	24.4

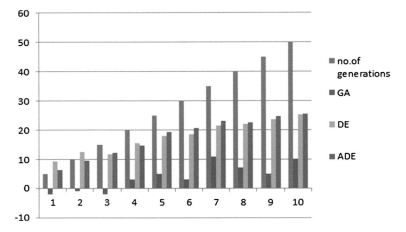

Fig. 6.3 Performance evaluation of GA, DE and ADE for Test Data Set III

Table 6.3 Fitness values of GA, DE and ADE for Test Data Set III

No. of generations	GA	DE	ADE
5	−2	9.2	6.2
10	−1	12.6	9.6
15	−2	11.8	12.2
20	3	15.6	14.6
25	5	18	19.4
30	3	18.6	20.8
35	11	21.6	23.2
40	7	22	22.6
45	5	23.6	24.6
50	10	25.4	25.6

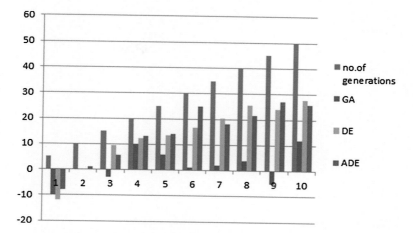

Fig. 6.4 Performance evaluation of GA, DE and ADE for Test Data Set IV

Table 6.4 Fitness values of GA, DE and ADE for Test Data Set IV	No. of generations	GA	DE	ADE
	5	−10	−11.8	−8
	10	0	0	1
	15	−3	9.6	5.8
	20	10	12.2	13.2
	25	6	13.4	14
	30	1	16.6	25
	35	2	20.4	18
	40	4	25.6	21.6
	45	−5	24	27
	50	12	27.6	26

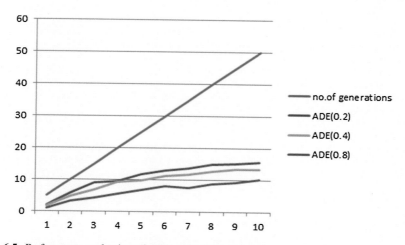

Fig. 6.5 Performance evaluation of ADE with different crossover rates

Table 6.5 Fitness values of ADE with different crossover rates

No. of generations	ADE (0.2)	ADE (0.4)	ADE (0.8)
5	1.98	1.82	1
10	5.8	4.86	3.3
15	9.02	6.76	4.26
20	9.62	9.26	5.54
25	11.84	9.66	6.8
30	13.14	11.32	8.04
35	13.68	11.84	7.56
40	15.16	12.74	8.9
45	15.32	13.56	9.32
50	15.7	13.46	10.34

6.5   Conclusion

This paper proposed a new mutation strategy for DE which is known as **"ADE/best-worst-rand/1"** and this is known as advanced differential evolutionary algorithm for pairwise sequence alignment. Various other methods are also used for sequence alignment like genetic algorithm, differential evolution apart from Advanced Differential Evolutionary algorithm. Better results are observed from ADE compared to genetic algorithm and differential evolution algorithm, which is being shown by the increasing fitness value with increase in number of generations. The average fitness value of advanced differential evolution is more than differential evolution and genetic algorithm. Applying ADE on Multiple Sequence Alignment is our future endeavour.

References

1. Batzoglou S (2006) The many faces of sequence alignment. Briefings Bioinform 6(1):6–22
2. Carrillo H, Lipman D (1988) The Multiple sequence alignment problem in biology. SIAM J Appl Math 48(5)
3. Soniya A et al (2015) A novel two-level particle swarm optimization approach for efficient multiple sequence alignment. Springer, Berlin
4. Shu J-J, Ouw LS (2004) Pairwise alignment of the DNA sequence using hypercomplex number representation. Bull Math Biol 66:1423–1438
5. Mount DW (2001) Bioinformatics: sequence and genome analysis. Cold Spring Harbor Laboratory Press
6. Benothman M et al (2008) Pairwise sequence alignment revisited-genetic algorithms and cosine functions. In: Proceedings of communications and information technology, Greece
7. Needleman SB, Wunsch CD (1970) A general method applicable to the search for similarities in the amino acid sequence of two proteins. J Mol Biol 48:443–453
8. Agrawal A, Huang X (2008) Pairwise DNA alignment with sequence specific transition-transversion ratio using multiple parameter sets. In: International conference on information technology ICIT'08, pp 89–93
9. Bellman RE (2003) Dynamic programming. Dover Publications, Dover

10. Agrawal A, Khaitan SK (2008) A new heuristic for multiple sequence alignment. In: IEEE international conference on electro/information technology, pp 215–217
11. Taneda A (2010) Multi-objective pairwise RNA sequence alignment. Oxford J Bioinform 26 (19):2383–2390
12. Feoktistov V (2006) Differential evolution: in search of solutions. Springer, Berlin
13. Gondro C, Kinghorn BP (2007) A simple genetic algorithm for multiple sequence alignment. Genet Mol Res 6(4):964–982
14. Notredame C, Higgins DG (1996) SAGA: sequence alignment by genetic algorithm. Nucleic Acids Res 24(8):1515–1524
15. Goldberg DE (1998) Genetic algorithms in search, optimization, and machine learning. Addison-Wesley, New York
16. Gondro C, Kinghorn BP (2007) A simple genetic algorithm for multiple sequence alignment. Genet Mol Res 6(4):964–982
17. Notredame C et al (1997) RAGA: RNA sequence alignment by genetic algorithm. Nucleic Acids Res 25(22):4570–4580
18. Price KV et al (2005) Differential evolution: a practical approach to global optimization. Springer, Berlin
19. Storn R (1996) On the usage of differential evolution for function optimization. In: Proceedings of biennial conference of the North American in fuzzy information processing society, pp 519–523
20. Kukkonen S et al (2007) Solving the molecular sequence alignment problem with generalized differential evolution 3 (GDE3). In: Proceedings of IEEE symposium on computational intelligence in multicriteria decision making, pp 302–309

Chapter 7
Swarm Intelligence and Variable Precision Rough Set Model: A Hybrid Approach for Classification

S. Surekha and G. Jaya Suma

Abstract Decision Tree is the most influential machine learning technique for the classification of real world datasets. Entropy based traditional Decision Tree classification methods often produce poor results due to the vagueness, uncertainty and noise in the data. In the past few decades, Rough Set Theory (RST) has been succeeded in dealing with uncertainty and vagueness. Thus, incorporating Rough Set concepts in the construction of Decision Tree for the classification of vague and uncertain data, yields better results. But RST based Decision trees sometimes classifies data too excessively and leads to a major problem called model Overfitting. This problem can be rectified by the Variable Precision Rough Set Theory (VPRST); which allows some level of uncertainty and misclassification in the construction of decision tree and classifies data more accurately. Even though the classification technique is very efficient, due to the increasing dimensionalities of the data sets the accuracy of the classification model is affected. So, from this high dimensional datasets with huge set of features, there is a need to select the most promising ones that contribute maximum for classification. In this paper, the most popular swarm intelligence technique namely Artificial Bee Colony algorithm is used for selecting the robust features and the reduced data set is submitted to the VPRST based Decision Tree for classification. In turn, a Reduced VPRSDT was obtained with promising results and it outperformed the traditional methods of tree induction, both in terms of reduced tree size and significant increase in accuracy.

Keywords Decision tree · Machine learning · Swarm intelligence · Artificial Bee Colony algorithm · Variable Precision Rough Set Theory

S. Surekha (✉) · G. Jaya Suma
JNTUK-UCEV, Vizianagaram, India
e-mail: surekha.cse@jntukucev.ac.in

G. Jaya Suma
e-mail: hod.it@jntukucev.ac.in

S. Surekha
Research Scholar, JNTUK, Kakinada, India

© The Author(s) 2016
P.V. Lakshmi et al. (eds.), *Computational Intelligence Techniques
in Health Care*, Springer Briefs in Forensic and Medical Bioinformatics,
DOI 10.1007/978-981-10-0308-0_7

7.1 Introduction

Classification is one of the most powerful data mining tasks, and can be applied successfully in many areas, especially in medical field. Medical field demands for efficient classification techniques to diagnose diseases with precise accuracy. Any misclassification of diseases can severely affects the patient's health condition and sometimes lead to other disorders also. But accurate diagnosis can be achieved only by building the best classification model through proper training. Now-a-days, machine learning techniques are being applied successfully in the medical field to develop classifiers for the diagnosis of diseases with precise accuracy. Decision Tree (DT) [1] is one of the most commonly used machine learning technique [2] for classification. In decision tree based classification, a tree like structure is used to build the model. In the process of inducing the decision tree, one of the attributes of the dataset is selected as the splitting node and this node selection is generally based on the statistical measures like Entropy and Information gain. But, the traditional method for attribute selection often fails because of the uncertainty in the given data. To deal with the uncertainty several set theories have been evolved, and during past decades RST [3] is one of the popular set theories that deals with uncertainty and vagueness, because it does not require any additional information to deal with uncertainties. So, the concept of RST has been applied for selecting the splitting attribute in decision tree based classification [4]. RST classifies the data more accurately and this accurate classification sometimes increases the size of the decision tree and this increased tree size causes a major problem called Model Over fitting. To address this problem of model over fitting, allow some degree of mis-classification in partitioning of data while inducing the decision tree. This mis-classification rate is introduced by Ziarko [5] as a new parameter in RST denoted with β and this β introduces two new concepts called as the Variable Precision Explicit region and Variable Precision Implicit regions, and this new RST is named as Variable Precision Rough Set Theory (VPRST) [6]. In VPRST based decision tree classification, the attribute with highest Explicit region is selected as the splitting attribute.

But, for accurate diagnosis of any disease, in addition to the blood test results and other symptoms, doctors often requires patient's history also. This causes the size of the medical datasets to be increased day by day. Processing of these high dimensional medical datasets requires a lot of resources which in turn increases the space and time complexities of the underlying algorithms. Thus, feature subset selection techniques [7] are to be applied to reduce the dimensionality of the datasets by selecting the most appropriate set of features that contribute maximum for a particular task. Hence, Feature Subset Selection (FSS) is becoming very important and is to be optimized. A number of swarm intelligence techniques have been evolved for solving optimization problems. In this paper, a novel optimization technique based on the collective intelligence of the honey bee swarms called the Artificial Bee Colony (ABC) optimization technique has been used for selecting the most promising features of the Thyroid disease dataset taken from the Irvine UCI machine learning repository. The feature subset generated by the ABC algorithm is submitted to the VPRST based decision tree classification for the diagnosis of Thyroid disorder and the performance of the Reduced VPRSDT classification technique is estimated using 10-fold cross validation test.

The rest of the paper is organized as follows: Initially, Sect. 7.2 discuses the works related to VPRS based classification of Decision Tree and various approaches of ABC Optimization technique. The methodology of the proposed work is given in Sects. 7.3 and 7.4 which present the experimental results and observations on the Thyroid dataset. The final Sect. 7.5 gives the conclusion.

7.2 Related Work

Mao et al. [8] proposed a new approach for building decision trees based on the concepts of Variable Precision Explicit and Implicit regions of the Variable Precision Rough Set Theory. The criteria for node selection is based on the VPRST Explicit regions, the attribute with highest Explicit region is selected as the splitting attribute and the results instantiated the feasibility of the new approach. Wei et al. [9] presented an approach for handling uncertainty in the selection of splitting attribute for constructing decision trees based on the VPRST, VPRST allows some degree of misclassification in partitioning of records, which is a kind of pre-pruning technique and thus reduced the size of the tree which improved the accuracy rate of the classification process when compared with the traditional Entropy based decision tree construction. Karaboga et al. [10] studied various evolutionary computation techniques and compared the standard version of ABC algorithm with the standard versions population-based techniques namely Genetic Algorithm, Differential Evolution, ES and Particle Swarm Optimization and the experiments were conducted for the same number of iterations and population size. The study shown that the ABC technique requires very less number of control parameters other than the population size and number of iterations and is simple and very flexible when compared to the other population-based optimization techniques. ABC algorithm can be efficiently applied for solving multidimensional models. Jia et al. [11] presented a binary approach of ABC called BitABC, which used bitwise operation for the movement of the employed and onlooker bees and compared the performance of binary ABC with the other variants of ABC, DisABC, normABC, BinABC and also with Genetic algorithm. The experimental results proved that the proposed BitABC performs better than the other variants of ABC. Schiezaro et al. [12] applied a binary approach of ABC algorithm for data feature selection on various datasets and the results showed that the ABC approach for feature selection reduced the number of features and the classification accuracy is superior to that of the original set of features. The performance of ABC is also compared with standard Genetic algorithm and other swarm intelligence techniques namely Particle Swarm Optimization and Ant Colony

Optimization and the results demonstrate that the binary approach of ABC produced better results for some datasets.

7.3 Methodology

The Reduced VPRSDT classification technique starts by applying discretization techniques on the data, and the discretized data is given as input to the ABC algorithm for generating the most relevant set of features. After the dimensionality has been reduced, the reduced features are submitted to the VPRSDT for classification which in turn produced a Reduced VPRSDT classifier and then compared the results of Reduced VPRSDT classifier with the traditional tree classifier C4.5 [13].

The methodology of the Reduced VPRSDT consists of four phases namely Discretization, FSS using ABC swarm intelligence technique, VPRST based Decision Tree classification, and Evaluation of Accuracy of the classification model and is shown in Fig. 7.1.

7.3.1 Phase I: Discretization

Many Medical datasets consists of various test results including blood tests, biopsies tests etc., and these results are sometimes continuous values. These continuous values often affect the accuracy of the decision tree classification by producing more number of branches for all possible values of the selected splitting attribute. So, to improve the performance of the decision tree classification, it is required to discretize all continuous values in the given dataset. Discrete values are easy to interpret than continuous values. Discretization [14] improves the prediction accuracy by reducing the data size and by making the learning process faster. In this paper, Equal width supervised discretization is applied on the submitted dataset to overcome the problem of model over fitting.

7.3.2 Phase II: Generate Feature Subset Using ABC Swarm Intelligence Technique

ABC is a popular swarm intelligence technique introduced by Dervis Karaboga in 2005, which models the collective intelligent behavior of real honey bee swarms. The minimal model of ABC algorithm [15] described by the Karaboga consists of three main components; Food sources, Employee bees and Unemployee bees. There

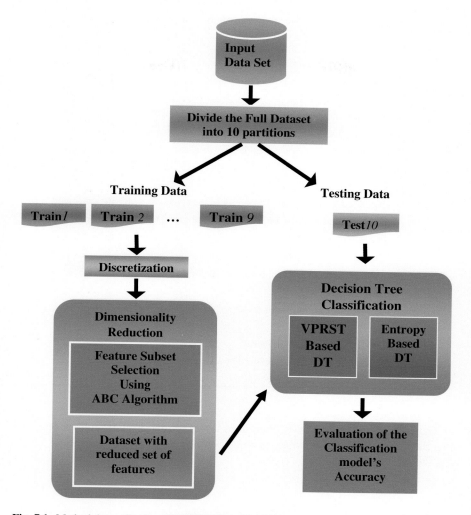

Fig. 7.1 Methodology of reduced VPRSDT for classification

are two kinds of unemployed bees; Onlooker bees search for food sources with the information gained from the Waggle Dance [10], whereas Scout Bees search for food sources randomly without any knowledge. The model of ABC algorithm proposed by the Karaboga, starts with an initial population of food sources and repeats the cycle of employed bees and unemployed bees for a predetermined number of times to come up with an optimal solution. In this paper, the binary approach of ABC algorithm has been implemented for generating minimal set of features. In binary approach of ABC algorithm, solution is represented by bit vectors [12].

Binary Approach of ABC Algorithm for Feature Subset Selection

Food Sources:
(1)   Generate initial population of N food sources,
       Where N corresponds to the total set of features in the submitted dataset

Employee Bee Phase:
(2)   Compute the individual's fitness by submitting the corresponding feature
       Subset to the classifier and consider the classifier's accuracy as the fitness value.

Onlooker Bee Phase:
(3) Repeat
(4)   Exploit each food source by determining the neighbors within the specified
           Neighborhood limit (NLIMIT) based on MR value. Generate a random number R
           Between 0 and 1. If R < MR, then the corresponding bit is set to 1 otherwise 0.
(5)   If the fitness of Neighbor is superior than the fitness of current food source then
(6)   Replace the Current food source with the Neighbor and Memorize the solution
(7)   Otherwise
(8)   Increment Neighborhood limit (NLIMIT)

Scout Bee Phase:
(9)   If NLIMIT > MAXLIMIT then
(10)   New food source is produced by setting the bit positions randomly.
(11) Until (maximum number of cycles)

The effectiveness of the ABC algorithm is determined by four control parameters; the size of the initial population, Neighborhood limit, Modification Rate Value and the number of iterations.

7.3.3   Phase III: VPRST Based DT Classification

Classification [16] is the most popular predictive data mining techniques used to predict the target class of an object on the basis of available information for a set of selected input attributes. Decision Tree is the most widely used machine learning classification technique, can handle high dimensional datasets with both categorical and numerical data and is very easy to understand. During the process of inducing a decision tree, appropriate attribute has to be selected as the splitting attribute at each level of the tree based on some standard measures. VPRST based approach of DT classification is based on the concepts of VPRST and selects the node with highest Variable Precision Explicit region as the splitting attribute.

Variable Precision Explicit Region:

Suppose KRS = (U, Q, V, ρ), be a Knowledge representation system [9] and the attributes set Q can be partitioned into subsets Q_C and Q_D, representing the set of conditional and decision attributes respectively. Let C and D represents two subsets such that $C \subseteq Q_C, D \subseteq Q_D$. $C^* = \{U_1, U_2, ..., U_n\}$ and $D^* = \{Y_1, Y_2, ..., Y_m\}$ are induced by $\tilde{C}$ and $\tilde{D}$, then

The variable precision explicit region is given in Eq. (7.1)

$$Exp_{C\beta}(D) = \bigcup\nolimits_{Y_i \in D} \underline{C}_\beta(Y_i) \tag{7.1}$$

where $\underline{C}_\beta(Y_i)$ is the β-lower approximation [9] of Y_i with respect to $\tilde{C}$.

With the introduction of the misclassification error **β**, the complexity of the tree can be reduced, because the β-lower approximation reduces the rigid requirements while partitioning inconsistent and noisy data, which further improvises the generalization ability of the Decision tree. The preferable values of β are in the range of 0–0.5.

In this RVPRSDT classification technique, the feature subset generated by the ABC optimization technique is submitted to the VPRSM based Decision Tree classification.

7.3.4 Phase IV: Evaluation of the Accuracy of VPRS Based DT Classification Technique

In this paper, 10-fold cross validation experiment has been conducted on the input dataset to estimate the generalization capability of the classification models. In 10-fold cross validation test, the entire dataset is partitioned into 10 equal sized partitions; among these 10 subsets, 9 subsets were used for training the model and the left over partition was used for testing the model, and this process is repeated for 10 times by shuffling the partitions. Then the performance of classification model is estimated in terms of its Accuracy. Accuracy of a classification model can be computed from the confusion matrix [17].

7.4 Experimental Analysis

Experiments were conducted on Thyroid disease dataset, taken from Irvine UCI repository [18]. The Thyroid disease data set is consisting of 498 records with 28 features and no missing values. These features include various symptoms and blood

Table 7.1 Description of the Thyroid dataset

Number of records	498
Number of attributes	28
Number of discrete valued attributes	21
Number of continuous valued attributes	06
List of continuous valued attributes	Age, TSH, T3, TT4, T4U, FTI
Number of classes	03
Classes	Hypothyroid, hyperthyroid and negative

tests that are required for the diagnosis of Thyroid disease. The brief description of the Thyroid dataset is given in Table 7.1.

In decision tree classification technique, if the splitting attribute is a continuous valued attribute then it produce more number of branches at next level, and these number of branches in turn produces much more sub branches and finally increases the size of the tree, which causes Over fitting. So, to improve the efficiency of the classification technique, Equal Width Supervised discretization technique has been applied on the 6 continuous valued attributes in the Thyroid dataset. The results of the accuracies of VPRST based Decision Tree for different β values for a 10-fold cross validation test is given in Table 7.2 and the effect of discretization is given in Table 7.3.

Submit the discretized data to the ABC optimization algorithm by varying the Modification Rate value. For different values of M.R, algorithm generated different set of features and the optimized feature subsets are given in Table 7.4.

From the Feature subset generated by the ABC approach, remove the irrelevant features from the dataset and then submit the reduced dataset to VPRS based

Table 7.2 Results of VPRSDT on thyroid dataset for different β values

β value	0.2	0.25	0.3*	0.35	0.4	0.45
Leaves	10	09	07*	07	01	01
Size	14	12	09*	09	02	02
Accuracy	96.57	94.85	96.57*	96.57	20.00	20.00

*Better Performance

Table 7.3 Performance of discretization

Classification technique	C4.5		VPRSDT
	Undiscretized	Discretized	
Number of features	26	26	26
Number of leaves	7	14	7
Size of the tree	13	19	9
Accuracy (%)	95.17	95.97*	96.57*

*Better Performance

Table 7.4 Feature subset for various MR values

S. No	MR value	No. of reduced attributes	Feature subset
1	0.1	6	Onthyroxine, TSH, T3, TT4, T4U, FTI
2	0.2	12	Age, Sex, Onthyroxine, Queryon Thyroxine, Sick, Thyroid surgery, TSH measured, TSH, T3, TT4, T4U, FTI

Table 7.5 Performance of VPRSDT for the feature subset obtained for M.R. = 0.1

β	0.05	0.1*	0.15	0.2	0.25	0.3	0.35	0.4	0.45
Leaves	5	10*	10	10	9	7	7	1	2
Size	15	14*	14	14	12	9	9	2	3
Accuracy	95.83	97.14*	97.14	97.14	94.85	96.57	96.57	20.00	20.00

*Better Performance

Table 7.6 Performance of VPRSDT for the feature subset obtained for M.R. = 0.2

β	0.05	0.1*	0.15	0.2	0.25	0.3	0.35	0.4	0.45
Leaves	9	10*	10	10	10	7	7	1	1
Size	23	14*	14	14	14	9	9	2	2
Accuracy	96.42	97.14*	97.14	96.57	96.57	96.57	96.57	20.00	20.00

*Better Performance

Table 7.7 Performance of VPRSDT for Thyroid dataset

Classification technique	C4.5		VPRSDT	
	Unreduced data	Reduced data	Unreduced data	Reduced data
Number of features	26	06	26	06
Number of leaves	14	14	7	10
Size of the tree	19	19	9	14
Accuracy (%)	95.97	96.37	96.57*	97.14*

*Better Performance

decision tree classification for various values of β and the observed accuracies for the two subsets are given in Tables 7.5 and 7.6.

The performance of the VPRS based Decision Tree has been improved for the feature subset generated by the ABC optimization technique for an M.R value 0.1 with 6 features only.

The comparison of accuracies of the VPRSDT for original features and reduced set of features is given in Table 7.7.

The comparison of the accuracies of VPRST based DT and C4.5 is shown in Fig. 7.2. And the performance of the VPRST based DT on unreduced and reduced data is shown in Fig. 7.3.

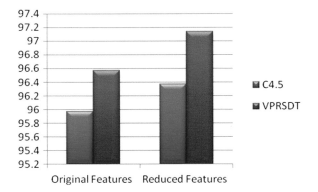

Fig. 7.2 Comparison of accuracies of C4.5 and VPRST based DT

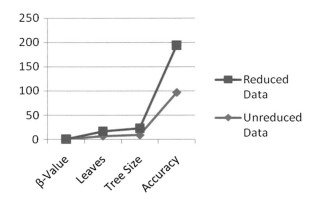

Fig. 7.3 Performances of VPRST based DT on unreduced and reduced data

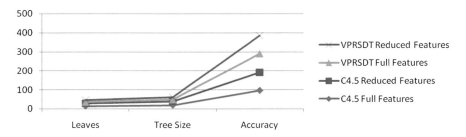

Fig. 7.4 Comparison of the performances of VPRST based DT and C4.5

The comparison of the performances of VPRST based DT and C4.5 for unre-duced and reduced datasets is shown in Fig. 7.4.

From the Fig. 7.4, it can be observed that the performance of the VPRST based Decision Tree is excellent for both unreduced and reduced data and its accuracy has been increased significantly for the reduced data.

7.5 Conclusion

In this paper, the reduced VPRSDT classification technique has been applied on the Thyroid disease dataset taken from the Irvine UCI repository. Thyroid dataset consists of 26 features and the observed accuracy of the VPRST based decision tree for the β value *0.3* and C4.5 is *96.57 %* and *95.97 %* respectively. The most influential ABC swarm intelligence technique has been applied for selecting the most appropriate set of features of the Thyroid disease dataset. Among the 26 features, the ABC optimization algorithm identified *6 features* as most promising features and when submitted these reduced 6 features to the traditional decision tree classification technique C4.5, the classification accuracy is observed as *96.37 %* with a tree size of *19* and when given to the VPTDT based DT the classification accuracy has been increased to *97.14 %* for the β value of *0.1* with a tree size of *14*. From the results, it can be observed that the efficacy of the VPRS based DT classification is better for both unreduced data (*96.57 %*) when compared to the accuracy of the traditional decision tree classification algorithm C4.5 on reduced dataset (*96.37 %*) as well as the for the reduced data also.

The Reduced VPRSDT classification technique, which is the combination of ABC optimization approach for feature subset selection and VPRS based DT classification, outperformed the traditional decision tree classification methods in all aspects.

References

1. Quinlan JR (1995) Introduction of decision trees. Alex Publishing Corporation, USA
2. Mitchell TM (1997) Machine learning. Mc Graw Hill, New York
3. Pawlak Z (1982) Rough sets. Int J Comput Inf Sci 11:341–356
4. Wei J (2003) Rough set based approach to selection of node. Int J comput Cogn 1(2):25–40
5. Ziarko W (1993) Variable precision roughset model. J Comput Syst Sci 46(1):39–59
6. Gong Z, Shi Z, Yao H (2012) Variable precision rough set model for incomplete information systems and its B-reducts. Comput Inform 31:1385–1399
7. Ladha L, Deepa T (2011) Feature selection methods and algorithms. Int J Comput Sci Eng (IJCSE) 3(5):1787–1797
8. Wei J-M, Wang M-Y, You J-P (2007) VPRSM based decision tree classifier. Comput Inform 26:663–677
9. Wei J-M, Wang S-Q, Wang M-Y, You J-P, Liu D-Y (2007) Rough set based approach for inducing decision trees. Knowl-Based Syst 20:695–702
10. Karaboga D, Akay B (2009) A comparitive study of ABC. Appl Math Comput 214(1):108–132

11. Jia D, Duan X, Khan MK (2014) Binary Artificial Bee Colony optimization using bitwise operation. Comput Ind Eng 76:360–365
12. Schiezaro M, Pedrini H (2013) Data feature selection based on Artificial Bee Colony algorithm. EURASIP J Image Video Process 2013:47
13. Quinlan JR (1993) C4.5: programs for machine learning. Morgan Kaufmann, CA
14. Liu H, Hussain F, Tan CL, Dash M (2002) Discretization: an enabling technique. Data Min Knowl Disc 6:393–423
15. Karaboga D, Basturk B (2007) A powerful and efficient algorithm for numerical function optimization: Artificial Bee Colony (ABC) algorithm. J Glob Optim 39:459–471
16. Bazan JG (2000) Rough set Algorithms in Classification Problem (Chapter 2). In: Rough set Methods and Applications. Physica-Verlag, Heidelberg
17. Son CS, Kim YN, Kim HS, Park HS, Kim MS (2012) Decision-making model for early diagnosis of congestive heart failure using rough set and decision tree approaches. J Biomed Inform 45:999–1008
18. https://archive.ics.uci.edu/ml/datasets/Thyroid+Disease

Chapter 8
Ayush to Kidney (AtoK) Data Science Model for Diagnosis and to Advice Through an Expert System

Kiran Kumar Reddi and Usha Rani Rella

Abstract In the present geographical conditions and increased urbanization, the soil and groundwater are contaminated with excessive fluorine solvents. The fluorinated water causes serious kidney problems, it's the need of the hour that man affected by kidney failures should be made aware of the reason behind the disease and how the proceedings of the life time could battle the failure environment of kidney, how much amount water and other intakes of food, body rest, etc. should be carried out for life time. The expertise data science model can perform and accomplish the task of suggestion and clears all the fears. Thus making the intelligent model incorporated by data science and expert systems are combined work for ayush to kidney. The data science models helps in diagnosis of kidney diseases, reason for failure and measures of cure through "Data Science Advisory System". We built an intelligent data science model to diagnose and advise the patient through expertise information about kidney life extension and its nurturing in a healthy way.

Keywords Machine learning · Support vector machine (SVM) · Data science · Self-organizing maps (SOM) · Expert system

8.1 Introduction

Our Vedas depicts that the five elements called pancha bhootas form the basic creation of universe. These are 'air' waste release by lungs, 'fire' waste through liver, 'water' waste through kidney. The kidney as an excretory organ serves the human system in performing the homeostasis. Nephron is the basic structural and functional unit of the kidney. Its chief function is to regulate the concentration of water and soluble substances like sodium salts by filtering the blood, reabsorbing what is needed and excreting the rest as urine. It has more than one million

K.K. Reddi (✉) · U.R. Rella
Department of Computer Science, Krishna University, Machilipatnam 521001,
Andhra Pradesh, India
e-mail: kirankreddi@gmail.com

© The Author(s) 2016
P.V. Lakshmi et al. (eds.), *Computational Intelligence Techniques in Health Care*, Springer Briefs in Forensic and Medical Bioinformatics,
DOI 10.1007/978-981-10-0308-0_8

nephron's that are on duty continuously. Manocha [1] experimentally proved that the mammal that take water containing 5 ppm fluoride shows Cyto-chemical characteristics that effects the kidney functionality and toxic addition to the body [1]. Fluoride intake crossing the optimal range 0.7–1.2 mg/L water causes renal structure change decreases to downfall tabular function and forms a primary lesion that leads to disturbance to water processing of the kidney [2]. The water processing of the nephron is given in Fig. 8.1 [3]. The functionalities disturbance is statistically analyzed and predicted the stones in kidney and paved away to move towards the machine learning techniques namely c4.5 algorithms, to improvise the prediction rate [4]. The analysis and work nature is carried out by Support Vector Machines in better accuracy rate and zero absolute error finding.

Fig. 8.1 Kidney functionality of Na^+ and H_2O excretion

Early risk AVF failure prediction through supervised techniques clearly given by Morteza research [5] in predicting the probability of complication in new hemodialysis patients AVF Arterio Venous Fistula (AVF) surgery.

The water processing of a nephron flow diagram gives a clear picture how necessary to the filtration blood and fluid contents of human body essentialities. The National Kidney Foundation in its 'Position Paper on Fluoride—1980' as well as the Kidney Health Australia express concern about fluoride retention in kidney patients. They caution physicians to monitor the fluoride intake of patients with advanced stages of kidney diseases [6]. The objective of this study is to extract the patterns of the kidney disease patient record and sample with new model of feature extraction using data science and soft computing techniques.

The second objective is to advise the patient for life time self-conscious of maintain food, life activities in a systematic manner that increase the ayush line of kidney. There is no effective evidence to summarize or suggestion fluoride water as a cause but it is reframed in 1998, as CKD Chronic Kidney Disease with fluorination effect. The question is whether the fluorinated drinking water causing CKD or not. In information visualization the Self-Organizing Map and SVM is an amazing tool [6]. Through mathematical analysis of the algorithm the fundamental implementation is simple and easy to understand the map allegory. In machine learning, SOM and SVM are supervised learning models with associated learning algorithms which examine the data and recognize patterns. Also, used for the analysis of classification and regression. The SOM and SVM training algorithm frames a model and manufacture it a non-probabilistic binary linear classifier.

8.2 Methodology

The data science is utilized to provide the e-patient record and remedies by AtoK model.

The basic contributions are

1. To detect the reason for kidney failure due to fluorosis in Krishna district.
2. To Analyze and estimate the severity of kidney failure through the data mining techniques namely support vector machine.
3. To frame the data of patient by an expert system, and advisory record using Soft Computing Tools.
4. This advisory record or information sheet should help the patient to follow himself in regard of food, intake of medicines, health consciousness maintenance etc.

Water soluble fluorine is not just hazardous but is dangerous for the next generation kidney functionality [7]. The identified information is the excessive levels of fluoride comprised normal water results in disturbance towards the normal functionality associated with critical parts of human body particularly the teeth, kidney, thyroid and bones. The optimal fluorine soluble should be in a range of 0.7–1.2 mg/L water. As the areas of Krishna district ground water is crossing these levels, the researchers have taken the position of nurturing palm through the soft computing—Data science, to build an intelligent model to "Diagnose And Advise" through educating the patient with hands on expertise information about kidney life extension and its nurturing in a healthy way. This is a serious problem because the shadows of the epidemic are viewed as chronic and the people are unable to accept even to get a female from the areas like Duganaputtaga of mandasa district to be their bridal partner. It's a small attempt to overcome this state for Krishna state in before [8, 9]. The earliest statements of media Mansingh Thanda (Krishna), November 4, 2014. Revised as the fluorides making the kidney health disturb.

8.3   Model Defined

In the process of identifying a positive decision we have moved with a novel combination of SVM along with Decision making System namely namely ID3 Algorithm. This particular work combination is followed in 4 phase manner.

1. Prediction Driven.
2. Model evaluation process.
3. Decision Interpretation model.
4. Impact of phases.

Figure 8.2 presents a data science model through combinations along with SVM decision tree ID3 Algorithm would give an intuitive exploration of strategies for decision making process. Based on above criteria, this model takes the job of prediction driven by support vector machines, Model evaluation process by ID3 decision support system, and the Decision Interpretation and diagnoses activity by expert system.

The Input data set forming with different attributes forms a separable leading symptoms and factors for the disease cause. The unnecessary limitations or so

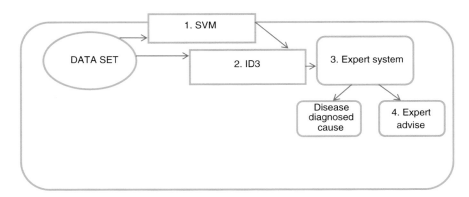

Fig. 8.2 Data science and expert system frame work designed

called uncertainty are covered-up or subject to avoidance [10]. This leads to a better decision making process. A novel combination of SVM with Decision Support System (DSS) is studied finally and presented in the proposed work. This combination follows three phased operations via. Predictor driven, agent evaluation and interpretation decision [11]. Such combination gives intuitive exploration and work out on uncertainty that occur before taking a decision.

This work has carried out the analysis and modification of existing models like SOMs, SVMs etc., and proposed their implementations for disease prediction. The existing algorithms have been suitably modified either in the form of preprocessing the inputs, deciding the optimal topology of the network or training the network in binary. There is not much work that has been done on CKD prediction using the kind of combination of algorithms proposed. Hence, it is felt that this is a solid contribution in the field of Data science prediction. Rise of kidney ailment cases worries Lambadi tribal of Krishna district.

8.3.1   AtoK Model Alias Computational Intelligence Model

This is the model ensemble to rejuvenate the kidney, and report produced by an expert system AtoK Model is one of the best estimate models to oversee errand movement in procedure of desire and reporting (expectation). It utilizes data on expense, and work execution to set up the present status of the task. By system for a preprocessing the data it permits the data appraisal to extrapolate current cases on the endeavor result. This is as for health information Analysis. The work plots the essential measures of the system and dissects its late modification gone for enhancing reliability in portraying task status, growing farsighted point of confinement, and considering the catastrophe control [12]. For this the taken stage is Computational Intelligence, a Statistical Learning figuring can see how to perform essential errands by summing up from frameworks [13, 14]. As more data finds the opportunity to be accessible, also yearning issues can be dealt with. Computational Intelligence expansive utilized as a bit of fragile preparing, making profitable machine learning applications obliges a worthy measure of "faint workmanship" that is precarious in examining samples [15]. The data science for health of kidney as prominent work in the field of intelligent expert system models striving to satisfy the process of identification domain and the attributes associated traverse towards disease specific system that advise the next coming days life of kidney [16]. The decision tree is usually a composition that also includes actual node, branch as well as leaf node. Every single central node refers to a analyze upon capability, just about every branch refers to the result connected with test and just about every leaf node supports the particular course tag. The particular topmost node from the tree could be the actual node. The decision tree tactic can be more robust pertaining to classification issues.

8.4 Conclusion

The data science expert system has the potential of decision making. The work is exposed as a nurturing palm to the diagnose and advisory activities by the effective utilization of statistical analysis and machine learning concepts. This model can perform diagnoses and also gives suggestions. This can be concluded as striving to educate that health of human is in the hands of expertise information and kidney life extension and its nurturing in a healthy way through the data science expertise machine learning.

References

1. Manocha SL et al (1975) Cytochemical response of kidney, liver and nervous system to fluoride ions in drinking water. Histochem J 7:343–355
2. Singh A et al (1963) Endemic fluorosis. Epidemiological, clinical and biochemical study of chronic fluoride intoxication in Punjab. Medicine 42:229–246
3. Turner RT et al (1989) The effects of fluoride on bone and implant histo morphometry in growing rats. J Bone Miner Res 4:477–484
4. Kaladhar DSVGK et al (2012) Statistical and data mining aspects on kidney stones: a systematic review and meta-analysis. 1:543. doi:10.4172/scientificreports.543
5. Rezapour M et al (2013) Implementation of predictive data mining techniques for identifying risk factors of early AVF failure in hemodialysis patients. Comput Math Methods Med Article ID 830745, 8 p
6. Naccarato A (2012) National Kidney Foundation (NKF), "The facts about chronic kidney disease (CKD)," National Kidney Foundation
7. Garcia JG et al (1991) Sodium fluoride induces phosphoinositide hydrolysis, Ca^{2+} mobilization, and prostacyclin synthesis in cultured human endothelium: further evidence for regulation by a pertussis toxin-insensitive guanine nucleotide-binding protein. Am J Respir Cell MolBiol 5(2):113–124
8. Wadhwani TK et al (1941) Pathological changes in the tissues of rats and monkeys in F toxicosis. J IISC 35(3, Sec A):223–230
9. Cittanova ML et al (1996) Fluoride ion toxicity in human kidney collecting duct cells. Anesthesiology 84(2):428–435
10. Chandrajith R et al (2011) Dose-dependent Na and Ca in fluoride-rich drinking water—another major cause of chronic renal failure in tropical arid regions. Sci Total Environ 409(4):671–675
11. Phillips PH et al (1934) Chronic toxicosis in dairy cows due to ingestion of fluorine. Univ Wisconsin AgriExp Stat Res Bull 123:1–30
12. Chandrajith R et al (2011) Chronic kidney diseases of uncertain etiology (CKDue) in Sri Lanka: geographic distribution and environmental implications. Environ Geochem Health 33(3):267–278
13. Elgendy N et al (2014) Big data analytics: a literature review paper, Research Gate, Sep 21
14. Taylor JM et al (1961) Toxic effects of fluoride on the rat kidney. Acute injury from single large doses. Toxicol Appl Pharmacol 3:278–289
15. Singla VP et al (1976) The kidneys. Fluoride 9:33–35
16. Kour K et al (1980) Histological findings in kidney of mice following sodium fluoride administration. Fluoride 13:163–167

Printed in the United States
By Bookmasters